ISLANDS OF GRASS

TREVOR HERRIOT

ISLANDS OF GRASS

PHOTOGRAPHS BY
BRANIMIR GJETVAJ

COTEAU BOOKS

© Trevor Herriot and Branimir Gjetvaj. 2017

All rights reserved. No part of this publication may be reproduced, stored in a retrieval system, or transmitted, in any form or by any means, without the prior written consent of the publisher or a licence from The Canadian Copyright Licensing Agency (Access Copyright). For an Access Copyright licence, visit www.accesscopyright.ca or call toll-free to 1-800-893-5777.

Edited by Bruce Rice and Joanne Havelock
Designed by Tania Craan
Printed and bound in Canada

Library and Archives Canada Cataloguing in Publication

Herriot, Trevor, author
    Islands of grass / Trevor Herriot ; photographs by Branimir Gjetvaj.

Issued in print and electronic formats.
ISBN 978-1-55050-931-1 (hardcover).--ISBN 978-1-55050-932-8 (PDF)

    1. Grasslands--Great Plains. 2. Grasslands--Great Plains--Pictorial works. I. Gjetvaj, Branimir, 1960-, photographer II. Title.

2517 Victoria Avenue
Regina, Saskatchewan
Canada S4P 0T2
www.coteaubooks.com

Available in Canada from:
Publishers Group Canada
2440 Viking Way
Richmond, British Columbia
Canada V6V 1N2

10 9 8 7 6 5 4 3 2 1

Coteau Books gratefully acknowledges the financial support of its publishing program by: the Saskatchewan Arts Board, The Canada Council for the Arts, the Government of Saskatchewan through Creative Saskatchewan, the City of Regina. We further acknowledge the [financial] support of the Government of Canada. Nous reconnaissons l'appui [financier] du gouvernement du Canada.

In memory of George Ledingham

Table of Contents

1	Prairie Eye	3
2	Gifts of the Prairie	27
3	More than Grass	51
4	Islanders: People of the Grassland	77
5	Possible Prairie	107

Appendix: A Call to Action   136
Notes   141
Acknowledgements   147
Photographic Information   149
About the Authors   151

. . . the uncountable souls of the grasses,
making one soul, one bending
at dawn. Where will my soul go
when it can't walk among them?

from "Desire" by Jan Zwicky,
in *The Long Walk*, U of R Press, 2016

1

It races ahead on the trail, tan bristles in its brush,
black-tipped, then dives into the wall of wheat,
something more dreamed than seen, imagined,
wondered—*cat, dog or fox?*

Prairie Eye

IT WAS ALONG THE NORTHERN EDGE of Old Wives Lake—a vast inland sea that year—where I am pretty sure I had the briefest glimpse of a swift fox. It zipped along the trail in front of the car but when I stopped to try for a photo it vanished into the cropland as I watched the wheat sway to its hidden passage.

If it was a swift fox—what else has a small fox body with a grizzled back and a black-tipped tail?—it was north of its core range, but there have been some reported sightings much farther north in recent years. After being driven from the land during the 20th century, through habitat loss, trapping and poisoning, the swift fox has become the most successful recovery story on the northern Great Plains. Thanks to the vision of Miles, Beryl and Clio Smeeton, and the more recent efforts of Environment Canada, the World Wildlife Fund and the Calgary Zoo, these cat-sized canids are once again back on the prairie and breeding on their own. A census done in 2005-2006 in Alberta, Saskatchewan and Montana counted 1,162 of them, predominantly born in the wild.

The blanket flower (*Gaillardia aristata*) is a hardy native that thrives in ditches and on even the poorest of prairie soils.

I failed to get an image of the fox that day, but like many people who spend time on native prairie, I take a camera with me when driving or walking through the landscape. Days later I look at the images on my computer and search, usually in vain, for the colours, lights, and shades I saw in the field. I have scores of photos of plains and buttes that overwhelmed me in person only to disappoint me in pixels. The results seldom match the moment they were meant to record, the scene or drama that moved across my retina and left a trace on my soul, but I keep trying anyway.

Fortunately, there are people who have worked hard to learn the trick of translating the unframed, encompassing world of prairie into two-dimensional frames that can be put onto a page and transport us from armchair back again to the domain of grass and sky. Branimir Gjetvaj is one of the best, not because he has the right equipment and technical skills—important as those are—but because he has come to know and love grassland places and has spent countless hours among the people and wild creatures who live there.

Mists rising from the floor of Saskatchewan's Frenchman River Valley in Grasslands National Park.

In the summer of 2015, Branimir and I travelled, together and independently, to some of the largest islands of native grassland remaining on the northern Great Plains, gathering images and stories that might bear witness to the beauty of this all-but-forgotten archipelago of endangered ecology at the heart of the continent. Some of the photographs in these pages come from those trips, but many others come from Branimir's own journeys, before and since, to Montana, Alberta and Saskatchewan.

The prairie world is solar, fired by a light that makes ranchers squint and painters mad, but the field and range it gives to the eye stirs something in our DNA. The late poet of the grassland, Bill Holm, spoke of its "horizontal grandeur." Reflecting on the Minnesota prairie where he lived, Holm said that the land "unfolds gradually, reveals itself a mile at a time, and only when you finish crossing it do you have any idea of what you've seen."

"There are two eyes in the human head," he said, "the eye of mystery, and the eye of harsh truth—the hidden and the open—the woods eye and the prairie eye. The prairie eye looks for distance, clarity and light; the woods eye for closeness, complexity and darkness.... One eye is not superior to the other, but they are different. To some degree, like male and female, darkness and light, they exist in all human heads, but one or the other seems dominant."[1]

I have wondered about that eye squinting at wide vistas beneath harsh light—where did it come from? How did it emerge from the shadowed realm of the trees?

Grasses are latecomers to the planet, having arrived on the scene only seventy-five million years ago, just after the first placental mammals took shape. Sparse at first, it took them another sixty million years before they came to dominate large expanses of land. Like another recent arrival, hominids, grasses have become very adept at colonizing even the most difficult habitats. And they are diverse. From bamboo to the bluegrass on suburban lawns, there are around 10,000 kinds of grass on the earth—a number on par with birds.

Some soil scientists believe that the arrival of grasses may have cooled our global climate, setting the stage for the burst of adaptation, abundance and variety of life we enjoy today in the late Holocene. Creating new sod and soils, grasslands changed

June grass (*Koeleria macrantha*) is a common native grass in the mixed grass and moist mixed grass prairie eco-regions.

Lightly grazed native grasslands support a diversity of flowers and grasses.

the climate by sequestering more soil carbon, reducing evapotranspiration, and reflecting more of the sun's energy back into space.[2]

It started with savanna, open grassy areas interspersed with trees—the acacia in Africa or oak and poplar here in North America. Co-evolving with new grazing species that could process grass, savanna grasslands took hold in large landscapes where mosaics of animal populations were scattered across geographically variable environments, providing the ideal conditions for an explosion of biodiversity in a cooler and drier biome new to planet earth.[3] And, among the species coming into the light of open grassy landscapes were the first upright hominin primates.

This sagebrush plain provides essential habitat for the greater sage-grouse and many other species that thrive only in large expanses of sage.

Around the time our early *Homo* ancestors show up in the fossil record, 2.0 to 1.8 million years ago, the environment was becoming even drier, creating more grassland in the African landscape. The level of herbivore energetic productivity in savanna is nearly three times that in woodlands, so suddenly there was a lot more protein and fat available for hunters, in the form of large herding mammals with the gut ecology required to process grass and convert it into flesh. Increased availability of certain fatty acids in turn was a necessary precondition for the rapid brain evolution that gave human beings the capacity to reason, reflect upon, and alter their environments.[4]

The prairie eye, an eye for spotting animals over wide vistas—the eye that appreciates the images in a book such as this—seems to have emerged as our bipedal ancestors responded to selection pressure favouring those who had brains with the expanded and reorganized sensory and motor control zones necessary to see far and make and throw weapons.

So where does that bring us today, in an era when our brains have led us to use more complicated tools to exploit the fertility in grassland?

No one had to establish climate targets for the grass, soil and grazing animals that helped cool the planet, setting the stage for modern biodiversity and ultimately the evolution of human consciousness. Unfortunately, over the last century, we have put that cooling influence at risk as we removed the carbon-sequestering, sun-reflecting grasses to produce high-yield crops in petroleum-intensive agricultural systems that annihilate biodiversity and create dead zones where the mouths of major rivers meet the sea.

When the first European settlers arrived on the North American Great Plains, roughly 360 million acres of shortgrass, tallgrass and mixed-grass prairie rippled through the heart of the continent from Canada to Mexico. Agricultural settlement and suburban development have reduced that figure to 70 million acres[5]—destroying more than 80 percent of one of the planet's great grassland biomes. By comparison, most estimates say the Amazon basin rainforest has lost approximately 20 percent of its tree cover.

As the International Union for the Conservation of Nature declared in 2008, "... temperate grasslands are now considered the most altered and beleaguered ecosystem on the planet."[6] Places that once held some of the greatest assemblages of wildlife have lost the most natural cover and receive the least protection: less than 3.9 percent of the earth's remaining temperate grasslands are officially protected.[7]

Winter or summer, prairie is a landscape that invites contemplation.

How did we come to such a pass? When did the prairie eye lose sight of unprocessed grassland terrain and settle for the simplified and systematized geometry of tame crops? What is it that has made it hard for us to see that natural grassland deserves as much protection and conservation as land covered by trees?

If there are answers to those questions they likely run back to our ancestors as well and the forces that turned grassland hunters and herders into grain farmers, but one way or another civilization has come into the 21st century labouring under the idea that land covered with grass is not complete until it has been turned over with a plow and made to produce corn, wheat, soy beans or canola.

The ethic was right there at the foundations of prairie settler culture, in the homesteading requirements. If you did not break enough of the land you were granted, the government denied your claim and would give it to the first person to build a shack and plow ten acres. In Canada, during settlement the Dominion Government gave ranchers 21-year leases on large tracts of land but the Minister of

Industrial agriculture, conventional and organic, has destroyed millions of acres of native grassland, in some areas eliminating more than 90% of the natural land cover.

the Interior could cancel any lease with two years notice, if he deemed that the land was needed for farming.[8]

Those of us who have come to privilege a generation or two downstream from our settler ancestors have some work to do: reconciling with the prairie and its original stewards dispossessed by colonization, yes, but also reconciling in our own hearts the split that has us on the one hand honouring our hard-working pioneer forebears and the dispossession and oppression they were often fleeing themselves, while fully addressing the injustice and disarray their arrival unleashed upon this land and its peoples.

Grassland wears that conflicted history on its sleeve. When you walk through prairie you can read the changes, see the remnants of cultures that have risen for a time and then faded or moved away. When I go now to walk the native grassland I know best—a patch in a valley complex an hour east of the city—I see the marks of a former world everywhere, the spoor of those who passed this way. In two miles

I walk by an archaeological dig, a former buffalo pound, burial sites, homesteader ruins, and broken down fences. If there have been heavy rains, arrowheads and bison teeth emerge on the trail.

For all that is missing, though, the grasses are still there. Sure, they struggle for sun and soil as brome and crested wheatgrass invade, but they are holding on. The birds may be dwindling year by year, but the songs they offer are the same ones they sang to the bison and those who hunted them. It is a small island of grass—a mix of private and community pasture perhaps seven thousand acres in all—but rich in its seeds, diversity, and soil ecology, its assets a forgotten treasure sown into the land like money in a mattress.

Should we ever, as citizens of this continent, discover that we all share an interest in the hidden bounty of our grasslands, we could halt the forces that continue to erode and degrade them.

That would take some collective power coming from moral high ground, to be sure. In fact, we have such a grounding in our history, a coming together that was intended to shape the way we use and share land. In the 1870s, settler and Indigenous people met to discuss and sign treaties: sacred agreements establishing the terms for settlement, but also recognizing the inherent rights of First Peoples whose original land-based ethics, cultures, and languages are as much a part of the grassland world as the song of the meadowlark or the swish of a bison's tail.

And now, a century and a half after Canada became a nation, there is a new international agreement that provides a much more recent commitment to the land-based rights of Indigenous peoples. In 2016, Canada signed and fully endorsed the UN Declaration on the Rights of Indigenous Peoples (UNDRIP). In articles 24-27, the declaration makes it clear that all nation states have an obligation to include Indigenous peoples in discussions and decisions that affect their rights to access and use "their lands, territories and resources, including those which were traditionally owned or otherwise occupied or used."9

The United States is the only nation on the planet that has not signed UNDRIP, but there will be continued pressure on American governments to fall in line with the new international order regarding Indigenous peoples and land.

Canada also signed the United Nations Convention on Biological Diversity (CBD) in 1992, releasing the Canadian Biodiversity Strategy in 1995. That national commitment was upgraded in 2010 when Canada adopted the CBD's Strategic Plan for Biodiversity, including the 2010 Aichi Biodiversity Targets. According to Target 18 of the accord, by the year 2020 Canadian governments will take measures to demonstrate respect for "the traditional knowledge, innovations and practices of indigenous and local communities relevant for the conservation and sustainable use of biodiversity, and their customary use of biological resources . . . subject to national legislation and relevant international obligations, and fully integrated and reflected in the implementation of the Convention with the full and effective participation of indigenous and local communities, at all relevant levels."[10]

The language in these treaties and international agreements may sound threatening to current land users, but the process of inviting Indigenous people into discussions of how we protect biodiversity in general and how we conserve and share public grasslands in particular has nothing to do with getting rid of cattle or ranchers.

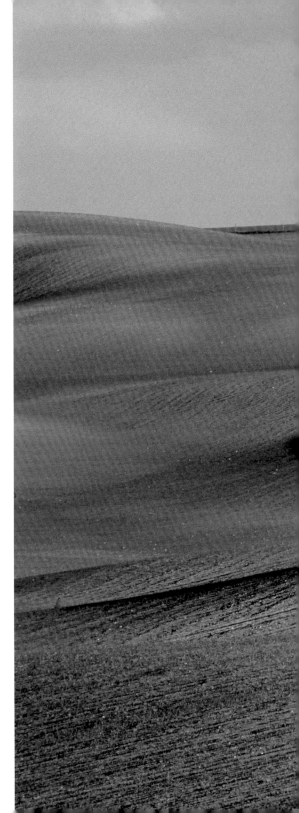

Subsidies and increasing economic incentives in recent decades have led farmers on the northern Great Plains to plow even poor soils in hilly terrain.

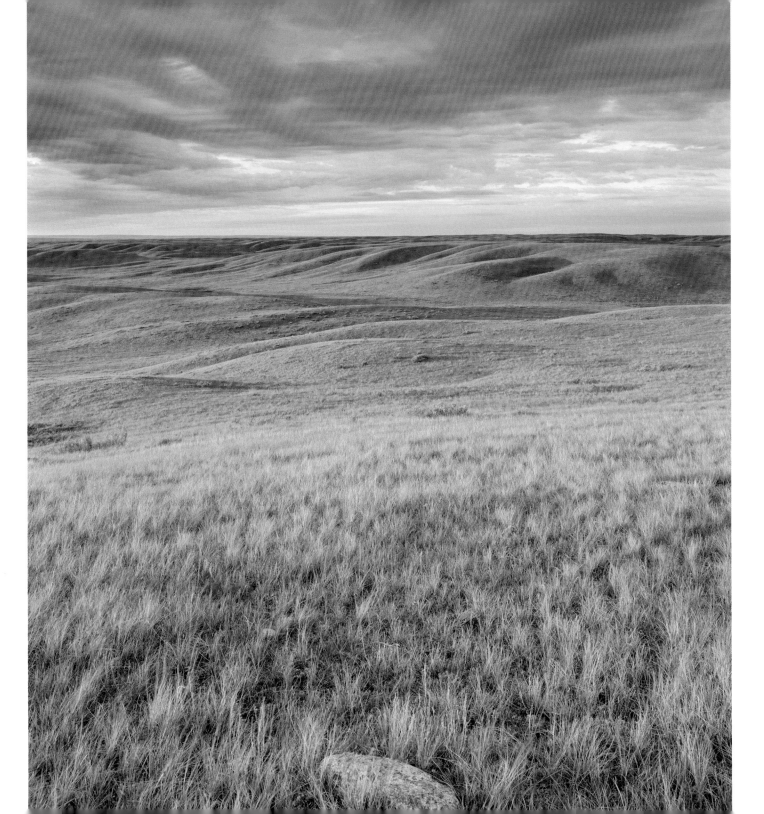

Anyone who looks closely at the forces threatening our private and public grasslands will recognize that the first line of defence has to be resident stewardship by people who have a cultural and economic connection to the land. On the northern Great Plains, ranching families on private and public lease lands and Indigenous peoples on tribal lands in the United States and reserve lands in Canada have been struggling against market forces and government policy to sustain that connection for more than a century.

Without their stewardship, our islands of grass would be even smaller than they are now. And cattle, when managed well, are a vital ecological substitute for wild grazers, one that sustains local rural economies and keeps the old-growth prairie intact and healthy.

Outside of the Prairie Farm Rehabilitation Administration (PFRA) community pastures, previously protected for more than seventy-five years by a program that was cut by the federal government in the spring of 2012, that legacy of protection has happened independently, and in spite of, government programming. While their neighbours took advantage of government incentives to plow their titled grassland and plant crops, ranch families managing vast expanses of private and publicly-owned native prairie have so far received almost no support from taxpayers or government programs to protect biodiversity. Granted, affordable grazing leases on government land have benefitted ranchers, but for more than a century, Canadian and American agricultural policy has favoured and subsidized the row-crop farmer and the rail transportation and export of grain, providing what economists call a "perverse incentive" for people to destroy their prairie and plant grain, oil or pulse crops.

During the worst decade of subsidization, the 1990s, taxpayers in Canada were paying more money on grain support programs than the agriculture industry itself earned.[11][12] Many of the Canadian subsidy programs ended with the century, but

The common "gopher" or Richardson's ground squirrel, creates habitat and provides food for other species.

to this day, there are income tax and insurance programs that provide incentives to drop a plow into native grass, and next to nothing to encourage landowners and managers to keep it intact.

That has to change. Economically sustainable grazing management as practiced by traditional ranching families should and must be a vital part of retaining and restoring health to our grasslands. For that to happen, though, we will need agricultural policy that recognizes the cultural, economic and ecological value of native grassland, that roots out perverse incentives in the market and in tax and insurance provisions that urge people to destroy the ancient grassland and seed it to crops.

Instead of hidden subsidies that place native grassland at risk, we need regulations and legislation that protect these endangered ecosystems by working with ranchers who want to make a living by grazing native grass, then retire and pass on their land in good condition to someone who will continue the tradition.

The urgency is growing as land prices rise, making it harder to justify and finance raising cattle on any given quarter section, and harder for producers to pass on their land and stewardship traditions to their children. The average age of ranchers is hovering around sixty and as they face retirement, some inevitably give in to the temptation to sell their titled land to someone who wants to plant lentils or subdivide the land and develop it for rural estates. Their government lease land can't be plowed or parcelled out but can easily fall into the hands of a feedlot operation or someone else who stocks it heavily for a few years, degrading the ecosystem, and then moves on.

How much native grassland is at risk of being cultivated? Some estimates say that there are still millions of acres that could be turned under to grow crops. The risks rise every year as universities and agribusiness corporations develop crop varieties that can grow under drier conditions and on less fertile soil.

Beyond hilly terrain and river valleys, most of the grassland world has been lost to cultivation.

Time is running out, but it is not too late to embrace our prairie remnants as geographies that contain both our history and our possibility as people who live where grass wants to grow. More than "working landscapes," they are places where we can yet discover what it means to be prairie people who regard the land not as commodity but as a community to which we belong.

    Most of the largest remaining contiguous blocks of original prairie on the northern Great Plains are still publicly owned, though there is mounting pressure to privatize and sell off government lands on both sides of the international border.

The continental population of the pronghorn, once in the millions, is now estimated at 700,000.

OPPOSITE: Little bluestem grass

A quick look at satellite maps, however, will show many sharp lines where the native grass protected on state, federal, or provincial lands gives way to privately-owned land that is entirely plowed under. Over time, and in the absence of regulation to protect it, all privately owned land will eventually shift to a form of use that ensures the highest economic return. In a suburban zone, a private woodlot of old growth forest arrives at its economic climax by hosting condominiums or a strip mall. Here on the Great Plains, any privately owned grassland will sooner or later be cultivated unless the hills are too steep and rocky. And the hill and valley complexes that protect native grass are increasingly susceptible to the dollars and dreams of people who want to build a McMansion with twenty acres out back where they keep a horse no one rides. I wish that the rancher's love of native grass was enough to protect it and the creatures who depend on it, but their refusal to plow grassland eventually goes to the grave with them, and the next owner, or the one after that, may not share the same ethics. Once public grasslands are privatized there is nothing to stop them from being cultivated or subdivided for ranchettes. A Crown conservation easement provides a minimal form of protection, but provinces like Saskatchewan struggle to monitor and enforce any of their environmental regulations. It is a rare prairie government that will charge a landowner who illegally ditches, drains or plows habitat. Over time, and with enough pressure from markets, and farmers pleading hardship, Crown held conservation easements will be overturned by one court or another.

The public lands we share are the last of the commons that reflect at least the potential of honouring the spirit of our treaties. When governments privatize them, even as they check off their "Duty to Consult" obligations, they are perpetuating and updating the colonial abrogation of treaties. Indigenous people lose their access to traditional lands overnight, eroding their already limited opportunities to experience the world that puts metaphor and meaning into their stories and language. Public policy loses its capacity—admittedly, underemployed to date—to ensure that the land will be managed not merely for private economic gain but for biodiversity and traditional use of biological resources by Indigenous people.

The national grasslands, leased rangeland, community pastures, and reserve and tribal grasslands on both sides of the 49th parallel should be treasured as the jewels that they are, as centres of natural beauty, cultural renewal, carbon sequestration, biodiversity, sustainable agriculture, research and learning. For that to happen, though, people who do not own cattle will need more places where they are welcome to walk and experience the prairie without having to drive five or ten hours to our one and only national park in the grassland biome.

People will defend and speak for places that they know—rivers where they fish, coastlines where they swim and camp, a forest where they hike. Not enough of us,

Winter reveals a new kind of beauty to the "prairie eye."

Indigenous or settler, get out to see large expanses of native prairie. Those who live in prairie cities, and many of those living on farms, have never heard the song of a Sprague's pipit, never watched a ferruginous hawk circle into the clouds, never shared a hilltop with pronghorn antelope.

The images and essays in this book are an invitation. They are meant to entice you to take a second look at native grassland, the last large islands of grass surrounded by a sea of cultivation that continues to lap at their edges. Your prairie eye will get some good exercise, but with any luck it will call you out to the land where you can let the plants and creatures take the lead.

The changes needed to engage with and restore the continent's prairie heartland are sometimes sparked by individual visions. The swift fox was saved by the efforts of a few inspired people; one persistent biologist with a passion for prairie wildflowers led the thirty-year effort to establish Canada's first grassland national park. Eventually, the sparks caught fire. Others took up the cause, carried the dream forward, found ways for grazing and conservation to thrive together, protected a nest site or a den from cultivation, welcomed bison and ferrets back to the land.

There is much work to be done and we can't wait thirty years for it to happen. Hanging in the balance is a massive carbon sink waiting to be deployed and deepened, more than thirty species at risk in need of programming and protection, and long-standing Indigenous and settler connections to grassland that must be respected.

If there is hope for the grassland world, it rests in the possibility that we will wake up, see our responsibilities to one another and to the land, and set aside our differences long enough to be inspired by the spirit of the prairie and its many gifts.

Let's begin there.

2

> The word for what the sky brings, and the earth
> embraces and relinquishes, in the shadow
> of the continent's great backbone
> spoken into the land
> is *maskotēw*—la prairie.

Gifts of the Prairie

PRAIRIE IS A PLACE where grass wants to grow more than trees or desert vegetation do, because of climate and soil. The grasslands of North America have been here for hundreds of thousands of years, and are as ancient as any old-growth forest, every bit as irreplaceable, every bit as venerable, worthy and rich with life. We think of grassland as a thin band of vegetation from just above ground level to the tips of the grass blades, a pelt of narrow-leaved plant growth covering plains and hills mostly in the Western half of the continent. But it is more than that.

Here in the Rocky Mountain rain shadow, grass is the organizing interface between things, a kind of ecological membrane that governs what is happening in the soil and water and on the surface. A healthy grassland strikes a shifting, dynamic balance between the water it stores in the ground and the water it stores in wetlands, between the resources it keeps in the soil and those it relinquishes to the web of life it feeds.

Depending on the timing and degree of grazing and fire, the nature of the soil and climate, grass organizes a diversity of ecotypes that shift over time and from place to place, providing organisms in the soil, on the land and in wetlands with particular niches that come and go on spontaneous schedules: several hundred species of grasses and flowering plants, and an astonishing array of animal life.

At the surface, below the leaves of grass, life gathers to a critical but mostly hidden threshold between upperworld and underworld. To see it you have to stop and get down on your knees and hold the taller vegetation aside with your hands. Biologists call it the cryptogamic crust, a miniature and fragile ecosystem of bacteria, algae, lichens, mosses, fungi and liverworts that grabs crucial nitrogen from the air and makes it available to roots below, that binds the soil together, protecting it from wind and water and keeping out weeds. On some sites this living crust may have thirty or more different species. Small, inconspicuous, but vital, these are the galaxies of life we stride over as we walk through native grass.

Below that threshold, more wonders live in the soil. Although grass provides the mechanisms that capture carbon and nutrients, the job of storing it happens underground.

Most of the carbon held in the ecosphere is found in soils. Unbroken native prairie sequesters a vast deposit of soil carbon. In his post-doctorate work at the University of Alberta, Professor Daniel Hewins of Rhode Island College estimates that while temperate grasslands make up only eight percent of the earth's surface they hold an estimated 300 gigatons of carbon.[13] However, when prairie is broken, soil bacteria convert the carbon into CO_2, which then heads into the atmosphere, contributing to global warming.

It has been said that there are more organisms living in a teaspoon of healthy soil than there are people on the planet. Grassland's primary gift is this ability to make soil filled with an astounding biodiversity of microbial life that captures and stores carbon and nutrients in a world beneath our feet which we never see and are just beginning to understand. From that foundation, grassland distributes its great wealth of fertility upward through the lives of the plants, insects, reptiles, amphibians, mammals and birds.

For grassland to continue working its miracles—building soil, storing carbon, capturing rainwater and storing it in the ground and in wetlands, providing diverse niches for other organisms—grazing animals must participate in the processes. However, it can't be just any grazing. The frequency and intensity of the grazing has

A biocrust or "cryptogamic" layer protects the surface of the prairie and the vital transition from life in the soil below to the life above.

Cattle ranching culture is providing an ecological substitute for the disturbance once provided by natural grazers.

to follow the plan set out in nature or else the soil actually loses much of its humus and the grass loses some of its capacity to be that vital membrane governing life.

Grazed at the right interval and intensity, the grass sloughs off old roots, and stimulates the growth of new ones, reaching deeper into the soil horizon to lift trace minerals upward where they can enter the food chain, moving from soil microbes to plants and the herbivorous insects and animals that depend on them. Each pat of dung from grazing animals in turn feeds the insect species and soil microbes that close the cycle by returning important elements to the system.

From this churn of nutrients on the prairie, those who eat high on the food chain—predators, including humans—benefit from the grazer's gift for turning grass into flesh. We don't have the internal ecology to make protein from grass, but cattle, bison and other ruminants do. The ecosystem inside the stomach or "rumen" of a ruminant—cow or bison—reflects and supports the health of the larger ecosystems outside.

Directly or indirectly, everything in grassland is fed by the grass. Whether you are a microbe or a buffalo, all flesh is grass. All of life is grass.

The peoples who have lived longest in grassland have always known this— that life is a gift that comes from grass, soil and rain. Among the gifts of the prairie, one of the most sacred for Indigenous people is sweetgrass (*Hierochloe odorata*). Its name speaks to its role in the spiritual traditions that run deepest in this part of the world. *Hieros* is Greek for "sacred" or "sanctified", and *Chloa* is Greek for "grass". One Blackfoot account of its origins says that Sweetgrass once lived in the sky along with another holy plant, the prairie turnip, which people today call Indian Breadroot (*Psoralea esculenta*).

We like to gather some Sweetgrass every summer and braid it up to hang in the rafters of our cabin on the edge of a prairie lake. Though it is locally common in our part of the prairie, it can be tricky to find until you get a feel for the way it looks. If I wait until it sets seed I can usually find a patch or two on the edge of the local bobolink pasture by watching for the sun to glint off the bronze florets swaying amidst sedges and wire rushes.

The nations who saw the holy in grass coexisted with grassland ecology for thousands of years until a new culture arrived—one whose peasant farmers had been growing crops for hundreds or thousands of years on the other side of the Atlantic. Displaced by war, persecution, overpopulation or an enclosure of the commons,

Prairie smoke, or three-flowered avens (*Geum triflorum*), lends a purple haze to grazed native prairie hillsides in spring.

OPPOSITE: Abandoned early tools of prairie "improvement."

Europeans drawn to the promise of free land with rich soils came to the Great Plains by the millions in a few decades spanning the turn from the nineteenth to the twentieth century.

Prairie homesteaders were told they had to make "improvements" on their land or else lose it—and the main improvement was done with a plow, although planting trees and erecting buildings were also important. They drove themselves with myths of land becoming more fertile if you plow it. These old beliefs linger in our land-use language: you "improve" a cattle pasture by scratching it and seeding non-native grasses, which are often invasive species. Land that hasn't been altered is "unimproved" or "wasteland".

Settler culture came of age with a version of history that has become a founding myth of our origins: there was nothing here but a few Indians chasing buffalo over treeless barrens. It was a wasteland, but with hard work and superior technology we have made the prairie into the breadbasket of the world. This, it seems, was the biggest improvement of all.

Myths have great power and this one—the lie of improvement—has squandered the gifts of the grass, of the soil life and fertility that has always been the source of health and well-being in grassland.

I have an old photo of my grandfather smiling ear-to-ear because he has one of the gifts in his arms. It was the fall of 1928, when he had a bumper crop after he plowed the native grass on his homestead on the edge of the Great Sandhills. The burst of nitrogen and other nutrients gave him a heavy crop for a couple of years, but he was rapidly using up the ancient fertility of the prairie. It would not be long before he and his neighbours would begin to learn the cost of plowing light soils in a dry land.

And so my father's earliest memories are of dust storms. Born in 1931, the world he came into had a sun that would disappear into blackness at midday. He remembers the angry look of the sky and the sound of sand scratching at the windowpanes. At meal times, his mother placed a cloth over the table and then quickly slipped plates of food beneath. The seven of them would reach under the cloth for something to eat and put it in their mouths before it could be covered in grit. By the time he was seven, his parents were joining the exodus of homesteaders abandoning their farms and moving to forested country. They pulled up stakes and headed north with their family, leaving exhausted soil that today's farmers resuscitate with artificial nitrogen.

Grasslands hold our history as prairie people, from ancient hunters to the Métis winterers to the first European settlers to homestead. They are among the last places where we can go to see what the prairie was like, the land that greeted our ancestors, Indigenous and settler alike.

The Saskatchewan Archaeological Society says that the province's remaining large pastures of native grass contain thousands of archaeological sites—tipi rings, burial sites, medicine wheels, bison kill sites—those known and many yet to be discovered. Keeping the grass intact and unplowed protects these features and any artifacts in the top layers of soil. Just knowing they are there, seeing a tipi ring or finding chert flakes on the surface, can bring you that much nearer to the life that thrived here for ten thousand years.

People who have lived on or near a large piece of native grassland will tell you that the land offers its stories to those who listen. Writers like Sharon Butala and Candace Savage have listened well, and with great respect and delicacy borne the stories of grassland into readers' hearts. Hannah Hinchman, a Montana artist and

The four white heart-shaped petals of the gumbo evening primrose (*Oenothera cespitosa*) open in late afternoon and remain open all night.

OPPOSITE: In spring, patches of native grassland shine with the blossoms of a native legume, buffalo bean (*Thermopsis rhombifolia*).

writer, says that to experience the land this way you have to go alone, that the spirit of a place retreats when you travel with other people. Alone, a walk draws you in and takes you down to another kind of awareness. The light seems different. The grass, different. You too seem more and less than yourself, a breath's free and passing gift away from who you were when you left home.

One of the loneliest jobs in grassland is the pasture riders'. Elise Dale travelled many hours on horseback checking on the cattle in Wolverine Community Pasture, near Lanigan, Saskatchewan—one of the first ten federal Prairie Farm Rehabilitation Administration (PFRA) community pastures to be cut loose from government support and leased to private management.

In the spring of 2012, Canada's federal government announced in Bill C-38 that it would disband the PFRA pasture program and turn most of the land back over to the prairie provinces. Saskatchewan had more than 60 pastures totaling 1.78 million acres of grassland, most of it native and rich with species at risk. The province announced immediately that it did not want to own or manage the pastures and began encouraging the grazing patrons to purchase them. To date, none have been sold. Instead, they are being leased out for private cattle management by former patrons who have had to form grazing corporations.

A couple of years into the transition process, Saskatchewan's Ministry of Agriculture decided to remove legislative protection from 1.8 million acres of Crown land that falls under the Wildlife Habitat Protection Act and offer it up for sale, auctioning it off to anyone in Canada who wants to buy land. Most recently, the Ministry announced that it will close its own community pasture system, reducing government oversight of these important ecosystems and placing an additional 780,000 acres of grassland at risk.

For nearly eighty years, the Canadian government managed some of Canada's last large expanses of native grassland under the Prairie Farm Rehabilitation Administration.

Oil and gas development on native grassland has become a major cause of its ecological degradation and fragmentation.

Despite requests from the grazing patrons and several conservation organizations, neither the Saskatchewan Ministry of Agriculture nor the Ministry of Environment have agreed to provide the kind of programming that is needed to manage the grazing with biodiversity in mind, to control invasive species and prevent oil and gas and gravel extraction activity from destroying habitat.

Concerned about the Wolverine pasture being closed and transferred into private management, Elise told me about her connection to the place.

She grew up on a mixed farm near the pasture where three generations of her

family brought cattle to graze. She has raised cattle herself but now focuses her time on breeding and training quarter horses. Riding through Wolverine, Elise said, she gained an appreciation for the history of the First Peoples who camped and hunted on the prairie. She talked about sitting on the north bank of Spoony Lake, where there is a teepee ring with a fire pit overlooking the landscape, with water and prairie running off to the horizon.

"Up on the buffalo jump, hunting blinds have weathered the years. Passing these structures, they always make me wonder to myself about the hunters that drew the

short stick and were planted up there to turn the herd of bison. Some of the blinds have been scattered from the elements and cattle rubbing on them, but one sits almost perfectly unscathed. I wonder, now more than ever, how many people will get to enjoy the very last of our wide open spaces and the story that this land tells."

Elise worried for the future of the pasture—that it might not be looked after the same way once the federal pasture manager, Eric Weisbeck, was gone. Eric, she said, was one of a new breed of conservationists. "... people who appreciate the land and the wildlife living off it.... He has done a great job of managing the grass and has made this pasture prosper for over a decade." To Elise, the Wolverine pasture's 16,000 acres not only fatten cattle but make for an important wildlife preserve—a haven for at-risk bird species, like the Sprague's pipit.

As someone who has lived in the country and raised livestock most of her life, Elise had trouble understanding the government's decision to cut the PFRA pasture program. "With an abundance of gravel in this pasture, local municipalities and provincial government agencies are trying to secure their piece of it. Steps need to be taken with proper conservation and archaeological easements, as well as mining regulations on gravel pits so this land will thrive under new owners and managers."

With demand for gravel on the rise, Elise fears that Wolverine's archaeological sites will not be properly protected. In the early stages of the transition to new management, it remains to be seen whether municipalities, the provincial government, or the grazing patrons will determine which, how, and when the gravel will be extracted. Gravel too is a gift of the prairie, but if extracting it is not weighed against the ecological and historic values of the lands, the demands coming from roadbuilding, urban development, and potash mines may overwhelm any thoughts of protection or mitigation.

Sometimes the history contained in an island of grass is only a generation or two away. One common story is the tragedy of families who were enticed into settling

Blue grama grass (*Bouteloua gracilis*) goes blond in the fall.

and cultivating land that should never have been broken. On certain PFRA pastures there are still the ruins of that sad chapter of our history—the leftover signs of the people who had to abandon or be relocated off the land in the '30s when the PFRA system was formed to conserve soil and the health of grazing lands.

Georgiaday Hall lives in Winnipeg, but her family's history is under the soil of the Swift Current-Webb PFRA pasture, west of Swift Current. A member of the Native Plant Society of Saskatchewan, Georgiaday traces her roots to a homestead contained in that pasture. And, like Elise Dale, she is worried about the future of the PFRA pastures.

Her grandfather, a widower, came to Canada, found his way west and then sent for his four young children in 1913. Buying a homestead in the sandy soils north of Webb, Saskatchewan, he planted his first crop in 1915. Georgiaday believes that it was wrong for the government to lure desperate immigrants out west and then give them such poor land to homestead, leading to disaster for her father's family and many more.

When her grandfather and many of his neighbours were moved off the land to make the community pasture in the 1930s, the family went broke. Georgiaday explains, "The children were hungry, shoeless, shabbily dressed and often cold. The police came and took my father and his brother. They were separated and my father was placed in the Children's Shelter in Regina."

From there Georgiaday's father, at age twelve, was turned out to work on farms as cheap labour.

On some of the PFRA pastures, farmers who had been enticed to settle on the drier regions of the prairie were removed to revert the land to Crown grassland.

"The farmers were supposed to make sure he got to go to school but he was kept at the farm to work. He was to receive a small wage and he did not. He ran away from four farmers in all. Sometimes the police caught him and returned him to the farmer and he was severely punished. His life was forever changed by not being allowed to attend school."

"Today there are oil wells on my grandfather's homestead. My father and his siblings were never compensated for the hardships and loss of family they were forced to endure."

Georgiaday believes that in honour of those settlers who were assigned unsuitable land and then later removed to make the pastures, the land should be managed and protected as native prairie.

"My father loved nature . . . and I appreciate all of the people who are trying to preserve land for the plants and animals we share this earth with."

The prairie coneflower (*Ratibida columnifera*) is a common sight on native prairie in July.

In a dry landscape like the Webb region where Geogiaday's grandfather homesteaded, all water, above and below the surface, comes as a gift. Lack of water may make a prairie, but the water it does retain is critical for the whole landscape's ecological well-being. Remove too much of the native cover from the land and the rain and spring meltwater does not infiltrate the ground properly. On land that has little perennial grass to hold the topsoil, a heavy rain will send water, soil and nutrients rushing into creeks and rivers, leading to downstream floods and algae blooms in lakes that collect waterborne sediments.

When healthy, well-managed grassland covers the land, every living thing in the watershed benefits. Sediments are retained, wild plants and animals thrive, ground water is recharged, and floodplains and watercourses can handle the water that

comes their way. If managers take measures to protect creeks and waterways from cattle damage, native grassland can provide vital source water protection with very little contamination from agricultural chemicals, and cleaner water in general.

Our brains benefit too. Psychologists are demonstrating that interchange with nature is necessary for the health and development of both our bodies and minds—in particular that we perform better mentally when we have regular contact with natural landscapes. Healthy, biodiverse and natural landscapes like prairies help keep us healthier—lowering blood pressure, improving immune cell activity and increasing our capacity for attention and creative problem solving.

A child, though, does not need words like "biodiversity" or "ecology" to be held and healed in the arms of nature. Katherine Arbuthnott, a Professor of Psychology at the University of Regina, has studied the links between time in diverse natural environments and cognitive and emotional health. It was in part memories from her own childhood that inspired her interest in the topic and her support for organized efforts to preserve grasslands.

When Katherine was ten years old her mother died. It was October. Through a long and sad winter, she would curl up with books, something she had always done with her mother. When school was out the next summer, and for every summer after that, she and her two younger brothers spent the holidays with her aunt and uncle at their farm.

"Aunt Myrtle and Uncle Alex farmed near Paswegan, Saskatchewan, and had four children, three girls and one boy, all older than me. There I was, a grieving city kid, used to being the oldest child and only daughter, suddenly becoming the youngest girl and just one of the 'farm kids' (which meant endless work, given the marginality of their farm and the number of mouths to feed) . . . that first summer was very difficult. I was grieving, in a context that was very strange to me and often overwhelming.

Moss phlox (*Phlox hoodii*) is one of the first plants to bloom in spring on native prairie.

OPPOSITE: Wetlands, small and large, form important centres of biodiversity in native grassland.

"And that is how I found the pastures. One day, when I was feeling particularly bereft and overwhelmed, I snuck away from the garden (where I was supposed to be weeding) to the neighbouring pasture of tall grass. I lay down on my back, watching the grass above me ripple in the wind and the clouds float lazily across the sky. Although I felt guilty for leaving my chores, I also felt comforted and peaceful, perhaps for the first time in many months. I stayed there until I started to hear the worried calls of my aunt and cousins in the farmyard behind me. I was scolded that day, but I'm sure my aunt recognized my need, because she never again mentioned my unexplained absences, although I often retreated to that field and those grasses both that summer and the ones that followed. I also learned to volunteer to bring the cows in for milking, which involved a long walk through other fields and pastures.

"In many real ways, that field and those grasses helped that grieving and overwhelmed little girl find Mother Earth, and a home that didn't shift, no matter the context in which I found myself. This is how the grasslands became home to me. I grew up, founded my own family of five, and seldom returned to those peaceful moments laying in tall grass and watching the sky. But nonetheless, these grasslands are part of my being, foundations upon which I live, a legacy of those days when they soothed a grieving child."

We need the natural grasslands and wetlands of the prairie as places where we can take refuge and restore our spirits. Try it the next time you are struggling with a problem, facing the imponderables of life or simply feeling blue. Find a stand of willows along a prairie river and look out at the floodplain, sit in the oaks and maples on the rim of a wide prairie valley, or in the edge of an aspen bluff where the prairie takes over. Imagine ancestors living in that transition between the refuge of trees and the prospect of a grassy plain with an abundance of bison or antelope.

Where the trees leave off and the grass begins there is an ecotone of human nature that reminds us that we have been shaped by forces—soil, grass and grazers—gifts of the prairie that even now can stir our souls, feed our consciousness, and teach us how to restore the earth.

American avocets lend their grace and colour to wetlands throughout much of the Great Plains.

3

Walk through prairie and you are
the summit, the perpendicular, upright
against all horizons, a vertical glyph
sliding over the level plain. Ride horseback, people say,
and you are less an intrusion—
the antelope and the hawk, unalarmed,
let you pass by.

More than Grass

IF YOU ARE LIKE ME, AND YOUR EQUESTRIAN SKILLS peaked at the fairgrounds pony ride, you might want to try a canoe or kayak.

I have paddled down the Souris River to pastures with field sparrows, lazuli buntings and snapping turtles, canoed through the eroding canyons of the South Saskatchewan River, where golden eagles and prairie falcons nest on buttes, and along the oxbows of the upper Qu'Appelle where pipits and godwits call from the valley slopes. But the Missouri River in Montana is hard to match for range of scenery, history and prairie life.

The first time I floated down the Missouri it was along the Lewis and Clark river route with a group of friends. One night we camped beneath a riverside grove of plains cottonwood trees. In the morning, I woke to a series of sharp barks and whistles that sounded like no bird or mammal I knew. I crawled out of the tent to investigate and, making my way through the cottonwoods, came to a broad floodplain between the river and the distant hills. The flats were grazed down as smooth as a billiard table but hundreds of small mounds were scattered across the surface. Each of these had a prairie dog, or a whole family, some standing peg-upright, others down on all fours with black tails flicking.

On his historic 1804 trip west, Meriwether Lewis wrote "this wild dog of the prairie... occurs in infinite numbers." Someone on the expedition decided it would be a fine idea to catch one and deliver it live to their benefactor, President Jefferson. After a lot of digging of mounds and hauling of water they managed to trap a prairie dog, and then shipped

it from Fort Mandan to Washington along with four live magpies.[14] Opening the container bearing these strange, and no doubt starved and bedraggled, citizens of the West, Jefferson must have paused to reflect on the wisdom of the Louisiana Purchase.

Ernest Thompson Seton, an early Manitoba naturalist and writer who went on to become the darling of New York society before reinventing himself again in New Mexico, once claimed that the North American prairies at the end of the nineteenth century held five billion black-tailed prairie dogs west of the 100th Meridian. Never afraid to peg numbers without benefit of survey data, Seton more famously put bison at 75 million before 1800. Regardless of the accuracy of that oft-quoted figure, bison were all but gone by the time Seton was making his guesses. Today we see prairie dogs taking advantage of bison-grazed pastures, but one has to wonder if the abundance of prairie dogs Seton witnessed had something to do with the sudden absence of bison. No one can say definitively whether removing North America's largest wild grazers, and then eliminating wolves and other predators toward the end of the nineteenth century, increased or decreased prairie dog populations during the early settlement phase that followed the demise of the free-ranging bison herds.

Like bison, though, the prairie dog once played a key role in plains ecology, both as food and in creating distinctive, short-cropped habitats for a community of other species. Although it now occupies a scant 2 percent of its former range, and even that with some help from conservationists and sympathetic ranchers, a key remnant population at the northern limits of its range has held on through more than a century of persecution and habitat loss.

Just above the Canada-U.S. border, at 49.20°N, 107.56°W, the black-tailed prairie dog persists on the floodplains of one stretch of the Frenchman River valley. The earliest "Frenchmen" of the valley were almost certainly Red River Métis families escaping the advance of colonization from the east. Just south of the 49th Parallel they followed some of the last bison on the northern Great Plains and set up wintering camps.[15] Their history is sometimes forgotten in the telling of how the region came to be settled, but a good portion of the French and Roman Catholic culture of the Val Marie area (named for the Virgin Mary) owes its roots to those first Métis wintering families and the Oblate priests who travelled with them.

Over 38 years of ranching, local historian and rancher Lise Perrault and her husband Fernand raised cattle and nine children on their piece of the prairie near Val Marie. Lise knew the Métis story well, and until her death in 2015 at age 91, told it every chance she got. She loved to show Val Marie visitors the link between the last bison hunt on the plains and NHL hockey star Bryan Trottier. In 1885, at Buffalo Butte west of Val Marie, Trottier's Métis ancestors took part in what may well have been one of the last successful hunts north of the U.S. border.

While the Perraults and many ranchers in the region have always found ways to co-exist with and even protect rare prairie creatures, it was a grassland scientist who brought the plight of the Frenchman Valley's prairie dogs to national attention. A campaign that began in the 1950s with prairie dog protection became one man's

Everything a bison does on the prairie—travel, graze, wallow, defecate, and die—creates ecological opportunities for other plants and animals.

OPPOSITE: The black-tailed prairie dog has returned from the brink of extirpation in Canada.

Cushion milk-vetch (*Astragalus gilviflorus*). Several species of milk-vetch spangle the prairie with colours ranging from creamy white to purple and blue.

lifelong mission that would ultimately conserve grassland habitat for black-footed ferret, swift fox, plains bison, and many other endangered species.

More than fifty years ago, when no one else in urban Canada was thinking about the prairie dog, George Ledingham was at his desk writing editorials calling for their protection. By then he was a professor of biology at the University of Regina, but his life began on a farm just west of Moose Jaw, Saskatchewan, where he explored small pastures of native grass in the Thunder Creek drainage. The relationships he witnessed there between grasses, flowering plants, water, soil, birds, bugs and mammals led him to dedicate himself to biology and grassland conservation. He was particularly fond of the *Astragalus* or milk-vetch genus of grassland flowers—a native legume that he would have known from early childhood. During a 1965 sabbatical he travelled the planet collecting specimens of milk-vetch wherever he found them, bringing them back to store in his beloved herbarium at the University of Regina. One he discovered in Iran now bears his name, *Astragalus Ledinghamii*.

Over time the long view of grassy landscapes fostered in him the patience and abiding tenacity he would need to wear away the mountain of resistance facing anyone who wants to make a park out of native rangeland.

In the late 1950s, as president of the new Saskatchewan Natural History Society (SNHS), which he helped to found, Ledingham began urging policy makers to establish a new national grassland park in the Frenchman River valley. The board of the society initially proposed a park that would extend more than 2200 square miles, a grand vision they would soon have to pare back to 900. In the fall of '59, the society passed a resolution calling on the Tommy Douglas government to establish grassland reserves and protect the prairie dog.

The Province responded by promising to make a small preserve in the Frenchman Valley for the endangered mammal. Undaunted, a committee convened by Ledingham presented a brief arguing for the establishment of larger grassland preserves to protect rare plants and animals, including an international one along the U.S. border where the Frenchman River heads into Montana. The document mentions the prairie dog, sage grouse, horned toad and others, and is one of the first times a conservation group made the case for the reintroduction of bison and swift fox.

Five years would pass before the province finally turned over to the SNHS a 160-acre pasture of Crown land that contained one of Canada's last prairie dog colonies. For the next two decades, that postage stamp plot of land was the only piece of Canada set aside for the primary purpose of grassland conservation. With Ledingham and others at the SNHS patiently advancing the cause, it eventually became a cornerstone upon which Parks Canada would found Grasslands National Park.

The private and leased Crown lands that ranchers eventually sold to Parks Canada

A cairn marks the prairie dog sanctuary established by the Saskatchewan Natural History Society, which became the seed of Grasslands National Park.

Though there is greater tolerance today, some animals, such as the prairie rattlesnake and coyote, have long been persecuted in ranch country.

OPPOSITE: The coyote has survived more than a century of bounties and poisoning campaigns.

during the 1980s were just as important in assembling the park. Fernand and Lise Perrault's Rocking 4 Ranch, east of Val Marie, was one of these critical pieces. Lise was the park's most fervent defender through the early years when other ranchers were openly hostile to the idea of a grassland park.

For thirty years, the thought that scientists and nature lovers from the city wanted to take land away from grazing became a barrier to any plans to make a park. In June 1969 when the SNHS held its summer meet at Val Marie, naturalists from all over the province came to camp in the Frenchman Valley. They were hosted by ranchers David and Ruth Chandler, who were members of the society. Two years earlier David had erected a field stone cairn at the society's prairie dog sanctuary. Each night of the campout, however, in the early morning hours, a small group of local cowboys snuck into the site to overturn tents and lasso outhouses.

Behind the night raids by a few pranksters was a broader feeling of betrayal and dispossession in the local ranch community that urban naturalists have never really understood. Imagine that your grandparents and great-grandparents had established through grit and hard work a good life on native prairie. That foundation, you knew, was part of every breath of chinook air you drew when you went to put feed out for the heifers in February. From one year's branding to the next spring's calving, from the hockey rink to the Wood Mountain rodeo, it was a life you believed in and wanted to pass on to your own children. Happy to live and raise your family close to nature in a remote and forgotten region of the continent, you would have trouble seeing the sudden attention from city dwellers and governments as anything but a threat.

At public hearings local men and women stood and said that the prairie and its wildlife had survived because of, not in spite of, their presence on the land. A park would only bring tourists and researchers to a vulnerable landscape, disturbing animals that thrive on isolation and the kind of stewardship private ranching provides.

During the first twenty years of the national park's history we learned that the ranchers were right about one thing: if you remove grazing from grassland it will eventually lose some of its diversity and health. Today, Parks Canada uses both cattle and bison to ensure the land receives some grazing, and relations with the ranching community have warmed considerably, particularly as many families now work for the park.

The park has also helped to protect creatures that ranching culture has not been kind to. Some of the prairie's most characteristic and ecologically important players—western rattlesnake, badger, Richardson's ground squirrel, plains bison, black-tailed prairie dog, swift fox, and black-footed ferret—were not made welcome by cattle producers, though that seems to be changing too. As well, several of the birds that use native grassland managed for cattle production are declining and on their way out. Setting some native prairie aside to be managed for ecological priorities has offered biologists a different kind of laboratory in which to search for solutions.

Ledingham knew that and so finally, in 1989, thirty years after he and the SNHS proposed the idea, Canada established its first and only national park dedicated to the conservation of native grassland.

Since that time, Ledingham's original vision of not only helping the prairie dog, but "extending a lifeline to the ferret"[16] and reintroducing the plains bison and swift fox, has come to life in the Frenchman River Valley.

The pronghorn is a survivor of the late Pleistocene extinction of 10,000 years ago.

OPPOSITE: Little bluestem grass.

The number of prairie dog towns and their total acreage in Canada has doubled since SNHS established its tiny preserve. In 2009, 34 black-footed ferrets were released in Grasslands National Park. Since then, according to Grasslands' current superintendent Kevin Moore, "volunteer trackers have found three new litters—a third generation of ferrets, indicating that they are reproducing at the park." Though their numbers have declined with a recent dip in the prairie dog population, the ferret's main food source, related to an outbreak of sylvatic plague, Moore expects to release more ferrets when the prairie dog numbers bounce back.

The swift fox, a predator not much larger than a Chihuahua, has recovered spectacularly after a century of absence. Once driven near to extinction by settlers plowing land and setting out traps and poison to kill wolves and coyotes, this cat-like fox is once again in the park and breeding across the native grasslands of southern Alberta and Saskatchewan.

Perhaps the most stirring of all reintroductions, though, happened in 2005, a year before George Ledingham died at age 95, when 71 plains bison were brought to the West Block of Grasslands National Park. Ten years later, a herd of nearly four hundred bison now thrives in the park, which is expanding year by year as more ranchers sell their leased and private holdings to Parks Canada.

Not long ago, I sat at a safe distance and watched a young bull in the park tugging on a purple flower. It was a milk-vetch. Somewhere George Ledingham was smiling.

Whenever I watch bison I wonder what they know—the ones fenced on farms with foreign grass to eat, or their counterparts in parks and reserves who multiply in managed herds until it is time for a cull. Or the little band of wilder ones at Sturgeon River who graze on fat fescue grasses south of Waskesiu in Prince Albert National Park, but sneak out of the park regularly to taste the hay on offer in local farms until they are chased back to their federal allotment.

And what was stirring in the cerebral cortex of the matriarch who led her herd of one hundred outside the fence to roam farmland south of Tisdale for several weeks just before Christmas in the winter of 2015? She needed no license to use the prairie. The genetic legacy left by her freer ancestors was enough to send her venturing through snowdrifts and across highways. Was she looking for better grass? What does she know that grass managers will never really know?

We may not understand all of the ways that the plains bison once helped maintain the health of our grasslands, but it seems fair to guess the role it played was

Dung beetles help fertilize and sustain the prairie, breaking up and distributing balls of dung from grazing mammals.

OPPOSITE: Long-billed curlews and other prairie birds prey on dung beetles.

central and went well beyond the direct effects of grazing. How did their wallows, dimpling the plains with pockets of mud and dust, play into the long-term patterns of plant succession and micro-community? Was there anything about their droppings that the prairie has missed since they've been absent?

With more bison grazing North American lands than at any other time in the past 120 years, an estimated 450,000,[17] there are now some opportunities to look at these and other questions.

Scientists for the Wildlife Conservation Society are studying how bison collect and then redistribute the wealth of the prairie—by depositing nutrients in the form of urine and feces and shedding their wooly hair, coveted by nesting birds and mice for its insulating properties. The greatest restructuring of all may be when a bison dies or is killed by a predator. From coyotes and vultures to beetles, flies and soil microbes, everyone comes to the banquet offered in the carcass. In turn, birds, reptiles, and mammals that eat insects are drawn to a secondary feast.

As well, the horning and rubbing habits of bison help keep grassland free of woody and thorny vegetation. Over time, bison can shift habitats dominated by shrubs or cholla cactus and yucca toward a grass-dominant community. As the only large wallowing mammal on the continent, they can reduce erosion in riparian areas by altering the slope of cutbanks and thereby allowing grass and other vegetation to stabilize the soil. Their wallows also create distinctive plant communities, affect surface soil structure, and create micro-wetlands. In wet years, a wallow will fill with enough water to become a breeding and foraging pool for amphibians and insects, which in turn feed birds and other larger creatures.

In the summer of 2013, I had a chance to visit Grasslands National Park with the bison expert who oversaw the release of 71 plains bison into the park—the first to graze the Frenchman River Valley in more than a century. Wes Olson—well over six feet tall, drooping mustache, piercing eyes—has the bearing of a park warden from

central casting, but he is the real thing, a man shaped by a life outdoors in the great parks of Canada's west: Banff, Waterton Lakes, Elk Island, Prince Albert.

I was helping to guide some guests from Bird Studies Canada, including writers Margaret Atwood and Graeme Gibson, on a tour of the park and community pastures nearby. A small band of young bison bulls grazed in the background. We were sitting amid the June grass and speargrasses on the rim of a large coulee slicing into the park from the north when Wes began to talk to us about buffalo wallows, dung beetles, and bison hair. "Bison manure," he said, "feeds several kinds of dung beetles. You see them from time to time on native grass. A small ball of something is rolling away from you between the grass stems and then you realize there is a bug pushing it along." According to Wes, by moving, storing and eating dung the beetles help to break apart and distribute the nutrients and microorganisms that are fostered by bison digesting grass and eliminating waste in their dung. Prairie birds, from grassland sparrows to long-billed curlews, eat dung beetles. They benefit from bison in other ways too.

The Mormon metalmark butterfly, a species at risk in Canada, lays its eggs near few-flowered buckwheat plants found on eroded soils.

OPPOSITE: Sharp-tailed grouse dance each spring on grass clipped short by bison re-introduced to Grasslands National Park.

"Not long after the bison release," said Wes, "we began to find bison hair in the nests of ground-nesting birds." After the wool of muskox, bison hair is the warmest natural fibre available to nesting birds. "Plus, it repels water and gives off an odour that may mask the scent of a bird nest, helping to throw nest predators off the trail." Nest survival is a critical issue for the three out of four grassland bird species that have been declining for decades in North America.[18]

Now that Grasslands National Park's bison population has more than quadrupled, the benefits to grassland songbirds have to be weighed against the needs of other species at risk. The cows seem to like calving in the biggest piece of critical habitat for the Mormon metalmark, one of Canada's rarest butterflies. Whereas, in

the past, the random movements of bison over millions of square miles of grassland across a range of climate and topography would have fostered a variety of ecological communities to provide niches for all creatures, today, on a small island of grass like Grasslands National Park, the process of sorting out how bison affect an endangered butterfly's habitat has to be balanced with a lot of other ecological priorities.

    No matter how that balance is struck, there is solace in knowing that the sharp-tailed grouse in the park are now choosing to situate their mating leks on ground the bison have clipped short and the nests of pipits, sparrows and longspurs, so vulnerable to late snowstorms and heavy spring rains, are lined with a kind of wool that birds have been using as long as the prairie has been prairie.

Birds that need grassland to survive—such as the Baird's sparrow and chestnut-collared longspur (right)—have declined significantly since settlement and cultivation.

OPPOSITE: The loggerhead shrike, a threatened species in Canada, is a songbird that preys on large insects and small birds, amphibians and mammals.

Unlike forest songbirds, which typically choose a perch on a tree when they sing, many of the larks, longspurs, pipits and buntings that live in open grassland sing on the wing. They climb upward and then glide back down to earth with their wings spread, releasing their songs from the air, at various heights, subdividing the space above the grass the way boreal birds will parcel out the forest high and low.

An old memory: sunup on a July morning in a patch of grama grass on the northern edge of Brokenshell PFRA Pasture. A 23,000-acre piece of the basin that forms the upper Moose Jaw Creek watershed, it is one of the largest prairie remnants within an hour's drive of Regina. I had come to sit and draw birds, but found myself instead, for the very first time, surrounded by an aerial song and dance show. Above a moraine of clay, sand, and gravel dumped when the last glacier paused on its retreat north, I watched a dozen lark buntings and chestnut-collared longspurs flutter up and down to stake their claims on a ridge that, 12,000 years or so later, was providing just the right updraft for thirty feet of songflight.

Heavy rains in June had set the blooms of coneflower, silverleaf psoralea and gaillardia. The speargrass was bouncing green light back to the sun from the swale on my side of the ridge. From earth to sky it was all song, an improvised jubilation: the swishing notes of Sprague's pipits spiraling three hundred feet overhead, meadowlarks proclaiming from the fence posts, longspurs and buntings teetering in the breeze, and a dozen Baird's sparrows singing from who knows where. Something shifted and suddenly a group of longspurs was moving together in a loose string, as if small kites were being sent aloft and then allowed to sail back into the grass. The air resounded with their song in a music that is as suited to the prairie as the scent of sage or the shamble of a badger.

A newer memory: last summer on a July afternoon I found myself driving through farmland south of Highway 39 near the town of Lang not far from Brokenshell. In the middle of miles of cultivated land I stumbled on a native prairie remnant perhaps 640 acres or so in size—a small island but just big enough to host a mix of grassland birds.

There were no cattle in the pasture so I stepped over the fence and walked through the speargrass, sagebrush, and bluegrasses. Right away I began hearing the flight song of chestnut-collared longspurs, a pleasant tinkling sound. I spent a half hour or so wandering through the pasture and following the longspurs. Between songs, they would fly to a new spot riding roller coasters in the air like goldfinches, while uttering a soft, dry rattle, until it was time once again to pause and sing as they fluttered over the grass.

A Swainson's hawk perched on a fence post, watching my movements from a comfortable distance. Two or three Savannah sparrows buzzed from the far side of the pasture and a single Baird's sparrow placed its zip-zip-trill on the wind.

Making my way to the southern edge where some taller grasses were growing in a wet spot, I found a small colony of bobolink. The males flew low over the grass, holding their wings in a shallow V and then adding their strange clanking song to the mix—as though the meadow might be hiding a toy band with a couple of springs popped loose.

An upland sandpiper arrived from nowhere, landed on a fence post and then shifted to another one before it gave its distinctive wolf whistle song. While the rest of the birds using the small fragment of native prairie would have wintered in the southern U.S. or northern Mexico, the sandpiper, the bobolinks and Swainson's hawk travelled across continents to spend the summer here. Six thousand kilometers away, the fields of southwestern Brazil, and the pampas of Argentina, Uruguay, Bolivia and Paraguay, offer grasslands that are a perfect winter counterpart to the prairies of the northern plains.

Choosing the perils and vulnerability of a twice-yearly migration is part of the narrative that makes our birds into small feathered heroes. Well, some are small, like the bobolink, which weighs about a $1.25 in quarters. At the other end of the scale there are prairie birds whose very size may have worked against them in the early years of settlement.

One of the first birds to disappear from the grasslands of the northern plains was also its largest—the whooping crane. It has been gone from our prairie landscapes for so long that we are inclined to think of it as a northern species that merely passes through the plains. What many people don't realize is that large wetlands in the Aspen Parkland and mixed-grass prairies were once the stronghold of the whooping crane.

Travelling up the eastern Qu'Appelle Valley in July, 1858, colonial explorer Henry Youle Hind wrote, "[The] white or whooping crane (*grus Americana*) was first seen today. This beautiful bird is common in the Qu'Appelle Valley and in the Touchwood Hills range."[19]

Farmers, following the footsteps of Hind and other scouts for the colonial enterprise, arrived in the 1880s and began tearing up the ancient sods, draining wetlands and shooting birds large enough for the cooking pot.

Whooping cranes, never as abundant as their smaller sandhill cousins, began rapidly thinning out as their nesting areas were converted into wheat fields, but

Colour me gone: in shades of green and yellow, the yellow-bellied racer, a species at risk in Canada, disappears rapidly into the grass.

OPPOSITE: Though today regarded as a northern species, the whooping crane once bred at major wetlands across the northern Great Plains.

unregulated hunting and climate extremes may have been the cause of their complete disappearance from the northern prairie in the 1930s.[20]

In the fall of 1921, two men hunting geese in a place called Buffalo Coulee near the West Central Saskatchewan town of Kerrobert shot three of the last whooping cranes to breed on the Great Plains. Fifty years later a letter from one of the hunters was published in the Saskatchewan Natural History Society's journal, *The Blue Jay*, accompanied by a photograph of the dead cranes. "We really didn't know they were so scarce ... we gave one away and the people told us they cooked the bird for four hours, then it was so tough they gave it to the dog."

Buffalo Coulee, where three of the last whooping cranes to breed on the prairie were shot in 1921.

The story includes a comment from a farmer who said that the cranes had nested in Buffalo Coulee "every year prior to 1921."

From the perspective of today's conservation ethics, it is easy to pass judgement on early settlers, but this was a moment in North America's history when most people believed nature to be inexhaustible and governments were just beginning to enact laws and treaties to conserve birds. Only five years earlier, Canada and the U.S. signed a new Migratory Bird Convention, taking an important step toward protecting all migratory birds from unregulated hunting. Enforcement would fall to

game guardians, men appointed by the federal government to roam wide stretches of countryside policing the hunting and fishing practices of settlers.

Not long after the Buffalo Coulee incident, Saskatchewan's first Chief Game Guardian, Fred Bradshaw, received a report of nesting whooping cranes from another Kerrobert area observer, a man named Archie Smith who lived on the shores of a lake to the north and east of Buffalo Coulee. Known at the time as "Baliol Lake," it now appears on maps as "Shallow Lake," an alkali marsh of two to three thousand acres that is today completely surrounded by the native grasses of the Progress

Male greater sage-grouse compete to gain the right to mate with females at a "lek" or dancing ground (OPPOSITE).

Community Pasture (former federal PFRA pasture). Smith informed Bradshaw that in 1911 there were as many as 12 whooping cranes nesting at the lake.

In May 1922, the spring following the shooting, Bradshaw dispatched guardian Neil Gilmour to look for nests in the area. In May he travelled to the broad alkali marshes near Kerrobert and found one of the last groups of free-nesting whooping cranes known at that time (a second group nesting 150 miles southeast at Saskatchewan's Lucky Lake would later disappear as their habitat evaporated during the drought of the 1930s).[20]

Most of the lakes where whooping cranes made their last stand on the Great Plains were enclosed within remote islands of native grass. Bill Cholon of Luseland, born in the dustbowl of the 1930s, was too young to witness the birds himself, but recalls his father's story from the 1920s when he saw the big white cranes or "herons" that nested at a large wetland just east of Kerrobert. It's still known as "White Heron Lake" and is now contained in what was until recently the Mariposa PFRA pasture. When a prairie fire roared over the grassland near their homestead, the family fled to the nearest water—the eastern shore of the lake. As they waited for the fire to pass, the prairie blackening around them, Bill's father spotted whooping cranes in the marsh.

But it was Neil Gilmour's trip to the region in 1922 that would lead to three official nest records for whooping cranes on the prairie, including two at Shallow Lake, only a few miles west of White Heron. Gilmour's account[21] was written up with some enthusiasm by the great American bird encyclopedist, Arthur Cleveland Bent, recounting what he believed to be one of the most important nest records in his 21 volume, *Life Histories of North American Birds*.[22]

After spotting a pair of whooping cranes, it took Gilmour several hours of tromping the muddy margins of the lake before he came upon their nest in a small patch of

open water: "The nest resembled a half-submerged cock of hay, flat on top and completely surrounded by water. Carelessly on the top of this mass of grass [likely sedge], was deposited the two large brownish-buff coloured eggs, about four inches in length."

In the letter that appears in Bent he concludes with, "The distressed whooping cries of the birds as I tarried at the nest, I shall never forget. They were the most remarkable calls I had ever heard. They boomed out and floated over the marsh, until the whole air seemed to vibrate."

Gilmour located a second nest and then Fred Bradshaw found a third one at Kiyiu (Cree for "eagle") Lake south of the town of Plenty, thirty miles from Shallow Lake.

In a letter Archie Smith wrote to Bradshaw in 1932, he reported that he saw the last pair nest in the summer of 1928. They visited briefly in 1929, he said, but did not return in 1930.

When the Big Dry of the '30s hit, the lakes evaporated and blew away in white alkali dust storms. By 1941 the world was down to fifteen whooping cranes wintering on the Texas coast and no one had a clue where they nested. Finally, in June of

1954, a forester named George Wilson travelling by helicopter to investigate a fire in the north side of Wood Buffalo National Park, in the Northwest Territories, spotted a group of large white birds on the ground that he believed were whooping cranes.

In 2016, there were 329 cranes in the Wood Buffalo flock and they raised 45 young—the third highest in fifty years of monitoring and conservation. There is no good reason why the next fifty years of whooping crane conservation should not be aimed at re-establishing a secure breeding population on the northern grasslands where they belong. The marshes of Mariposa and Progress pastures—now managed privately, but still public lands—could once again welcome the wild cries that Gilmour heard a century ago.

Few people living today in grassland were alive when the last whooping cranes nested on the plains, but most of them know the other large birds that remain. The whole lot of prairie songbirds may blur into a streaky brown anonymity, but birds like the greater sage-grouse (or "sage hen"), the burrowing owl, and the long-billed curlew have ways of getting your attention.

> *A rancher is out checking fence along the river breaks of the Missouri Coteau, when the wind suddenly resolves into a low, melancholy whistle that breaks and rises, shifting into a bubbling, rapid set of notes. He sneaks a brief glance up to the horizon. Something big and agitated is flying on a trajectory that will intersect with his chest in about five seconds. Staring at its long down-curved bill wide open, he wonders for a moment if it will turn away before impact. It gives its cry once more and then veers just before he has to duck. He remembers something his mother told him—the curlew has its nest nearby and it's best we leave it in peace.*
>
> . . . .
>
> *She looked forward to the spring arrival of the howdy owls—that was what she called them. They were just there every summer from April to September. As much a part of the forty-acre horse pasture north of the barn as crocuses at Easter. There was that one that bobbed along the fence line like he was following her. He'd stop at a post and bow up and down a few times, almost enough to make her want to bow back at him. The funny clatter he made with his bill as he floated over the shortgrass. And then that time he glided down and took a cricket that jumped ahead of her as she walked out to put a bridle on the sorrel. One summer some students came to talk about the owls—burrowing owls, they said. They had a paper they wanted signed, an agreement they said would protect the owls—from what, who knows, but they left a sign to post at the gate.*

A common sight: an eastern kingbird chasing a raptor—here, though, it is a ferruginous hawk, listed as "threatened" in Canada.

OPPOSITE: The long-billed curlew may winter near ocean shores but breeds each summer on the high plains.

There were always owls until there weren't.

· · · · · · · · · ·

It was the sound of the sage flats in April—a low whistle then pop, like some kid pulling a finger out of his mouth, over and over again on clear calm mornings. Their dancing grounds were a mile away but back then she could hear the sound from the yard every morning. Once the frost was out of the ground, she would sneak out in her housecoat and rubber boots. From the crest of the hill she could just make them out, twenty or more in a circle beaten flat, just a bit of grass surrounded by blue sage. The roosters all puffed up posing at one another, tails all fanned out, and flopping their chest feathers out and down every time they made the sound.

There are still a few curlews, burrowing owls, and sage grouse on the northern plains. We know more now than we did when the last prairie-nesting whooping cranes disappeared. There are government agencies and non-governmental organizations at work trying to slow the decline of these and other characteristic creatures of the prairie. Meanwhile, public opinion polls indicate three in five Canadians (62 percent) feel that the federal government is doing too little to ensure the survival of Canada's species at risk and should strengthen and better enforce Canada's endangered species laws.[23]

But laws and conservation agencies can only do so much. If grassland ecosystems and species are to make it through this century, it will happen because the people who live and work on the prairie have been supported by an informed citizenry demanding government policy in agriculture and environment that builds on and expands cultures of stewardship and sustainable land practices within settler and Indigenous communities.

The burrowing owl has declined spectacularly in Canada in recent decades, where it is now facing possible extirpation.

4

Makíínimaa, píitaa, maatáásii
   Long-billed curlew, golden eagle, burrowing owl
soohkiisiimsstaan, maotoyóópan, tsikatsíí
   meadowlark, tall grass, grasshopper
— tssáókioopiiwa
   — *he is living on the prairie*

Blackfoot/English

Islanders
People of the Grassland

LANGUAGE IS THE HUMAN HOME, a dwelling place, John Berger said. And yet we forget the mystery and beauty in language, the way it touches the more-than-human world. The right word placed the right way on the tongue is a form of magic. It can take you nearer or farther away from the hills and rivers where you live, utter the difference between an embrace and a grab.

But translation strips a lot of meaning as the language of a culture embedded in the land encounters that of a culture newly arrived. When I hear someone speak the Niitsitapi (Blackfoot) names for prairie creatures—the swift fox, the crocus or rattlesnake—the best I can manage is a vague sense that the words contain or express a regard for and bond with those beings that is impossible to understand outside the lifeworld enclosed by that language.

Still, the effort to listen has its own rewards. I met Paulette Fox (her Niitsitapi name is Naatawawaohkaakii) at a gathering of Canadian and American prairie conservationists held on the western side of the Cypress Hills in the middle of winter. An environmental scientist who is also a member of the sacred Horn Society and a Beaver Bundle holder within the Blood Nation, Paulette works across cultural boundaries to make the connection between human communities and the health of the land. She was a guest speaker at the workshop, charged with the daunting task of introducing a room of non-Indigenous scientists, ranchers, and grassland advocates to the notion that restoring the health of our native grass is as much a cultural and spiritual question as it is an economic and ecological one.

Tall, poised and confident, Paulette has a way of holding a room with her gaze and the cadence of her speech. Most of the

The tipi remains a meaningful symbol for plains Indigenous people, speaking to their continued responsibilities and cultural embedding in the prairie world.

people listening to her that day had come with their own ideas about how to begin the difficult work of conserving grassland over large, border-crossing landscapes. Yet here was someone looking through a different lens at that task, an Indigenous lens that holds cultural integrity and ecological integrity together in focus.

She spoke about the Iinnii Initiative, in which the Blackfoot Confederacy has been working with the Wildlife Conservation Society in the United States to see what could be done to return the iinnii (bison) to Blackfoot territory.

Leaving the Cypress Hills to drive home, my head was swirling with thoughts of bison ranging from Chief Mountain out onto the plains and foothills in and around Glacier and Waterton National Parks—two key conservation areas that hold hands across the border, forming the montane ramparts that front that stretch of the Great Plains.

The dream of reintroducing wild bison is fraught with controversy, but in April 2016, the Iinnii Initiative took a cautious first step. Five years of dialogue with elders of the Blackfoot Confederacy culminated in the release of 88 plains bison yearlings onto native rangeland east of Browning, Montana. Paulette was there, part of the group conducting ceremony and singing traditional songs to welcome the bison home.

In time, the plan is for the bison, which came from Canada's Elk Island National Park, to become a source herd for further Iinnii Initiative restoration efforts on Blackfoot lands in Alberta and Montana. With bison already leaving Yellowstone National Park and entering privately managed pasture, the prospect of more bison on tribal lands or public lands is enough to start a war in ranch country.

Some ranchers will tell you that bison are disease-carrying fence wreckers, but if you sit down long enough to listen to their concerns they will eventually come round to talking about culture, tradition, and rights to the land. The ranching communities of Saskatchewan, Alberta, Montana and the Dakotas oppose the reintroduction of bison onto public lands because they see it as a direct threat to their way of life. To families that have been grazing the prairie since the bison and bison hunting cultures were removed, talk of restoration or reintroduction is about as welcome as, well, a buffalo in your front parlour.

"What's next? Dinosaurs?" one stock grower said, when I asked for his thoughts on bringing bison back to his corner of Saskatchewan. He believes that putting more

Stone rings and other archeological features attest to the ancient relationship between the prairie and its Indigenous peoples.

OPPOSITE: The traditional knowledge of long-standing ranch families who have leased public grasslands for generations can work hand-in-hand with conservation strategies.

bison back on public land would give radical anti-beef, pro-bison environmentalists a foothold that could eventually see the complete disappearance of cattle from federal, state and provincial lands on both sides of the border.

Today's western ranch culture, built on grazing allotments and leases passed on from one ranch generation to the next, is already vulnerable to global market forces and negative campaigns from vegetarian, climate change and animal rights groups. Bison taking over public grazing lands might be the final death blow.

But does it have to be one or the other? Is there room for both wild and domestic grazers on the prairie?

"Humans like to think black and white," Kevin Ellison of the World Wildlife Fund once said to me, "either bison or cattle but not both. It does not have to be that way."

Raising livestock on native rangeland is one of the continent's more sustainable forms of agriculture. The majority of beef's much-criticized carbon footprint and ecological disarray comes after the animals leave the grass and enter the feedlot-to-market section of the supply chain. Finish the cattle on grass, the way bison once were, and everything from human health values to greenhouse gas emissions gets better.

Biologists like Ellison, who study grassland ecology, will tell you that, in general, private range management for cattle production is a way to sustain the health of our remaining native grassland. I have spoken to scientists working for the Bureau of Land Management in Montana or the Canadian Wildlife Service and they agree that many, if not most, ranchers are doing a good job of maintaining the condition of rangeland, and providing an economic justification for keeping the prairie grass side up.

So, then what about the both/and approach?

It was not cattle production that was incompatible with wild bison as much as row-crop agriculture. During the 1860s and '70s, while there were still relatively large herds of wild bison, the first open-range cattle operations took hold on the northern plains. More than five million cattle were driven north each year to be fattened on the open prairie, and in the fall those ready for slaughter would be driven or sent by rail back south. From western Kansas, Nebraska, the Dakotas, Montana and Wyoming, to the Canadian territories that became Alberta and Saskatchewan, thousands of square miles of native grass were supporting Herefords and other hardy breeds of cattle.[24] Meanwhile, until the mid-1880s, as many as ten million wild bison continued their annual migrations through the same terrain.

The open-range operations outlasted the wild bison by a generation or two, making it well into the 20th century. One of the last was also the northernmost—the 150,000-acre Matador lease on the north side of the South Saskatchewan River, a few miles southwest of Beechy.

A member of the Lakota First Nation performs a traditional dance in Grasslands National Park.

OPPOSITE: A single pass through prairie is enough to leave a lasting mark

At branding time, ranch communities renew social connections that are a vital part of the sustainability of domestic grazing on native prairie.

OPPOSITE: Moving cattle between pastures.

A number of years ago, I was at a food security conference in Moose Jaw where I had been invited to talk about the importance of grass in transforming prairie agriculture toward a healthier engagement with the land. The attendees were mostly urban people working in health and anti-poverty sectors—people who are connecting the dots between food systems and socio-economic ramifications downstream. After one of the afternoon workshops I wandered out to look at the book table and started talking to a man who had every sign of being a cowman. He introduced himself as Ted Perrin. His land is on the north side of the South Saskatchewan River, near Beechy.

I must've said something about liking the prairie in that area, because the next thing Ted said was, "We can thank the Texans for it." He smiled, wondering if I knew what he was getting at. And then he began to tell me about the Matador.

A Scottish livestock conglomerate, the Matador Land and Cattle Company was, in its day, the largest cattle empire in the world, with pieces of the open range from Texas to Saskatchewan. Each spring, between 1904 and 1922, the company would ship 6,000 head of cattle from breeding ranches in Texas to the Missouri Coteau hills along the Saskatchewan River. They were shipped off to market in Chicago after a summer of finishing on native grass. At two cents per acre, the Matador leased its Saskatchewan range from the federal government for a total of $3000 each year.

Sheep grazing is used to control invasive plant species.

OPPOSITE: Vast community pastures, such as this one at Govenlock, Saskatchewan, are a faint echo of the open-range cattle operations of the past.

The skills, values, and traditions of ranching culture harken back to the days of the open range.

The pieces of it remaining in Saskatchewan include some of the northern Great Plains' best examples of well-managed rangeland—including the Perrin ranch, Castleland, named for the locally famous Sandcastle formation, and the Matador Community Pasture, where Ted was born.

Stan Graber, one of the last of the Matador's cowhands, told the story of the final Matador roundup in a series he published during the 1990s in a farm newspaper called *Grainews*. He was in his late 80s at the time, casting his thoughts back to the spring of 1921 when he and a few other cowhands drove 3,500 steers from the Matador's Coteau range down to the river at Saskatchewan Landing, across the river and then south to the company's holdings along the Milk River in Montana, 350 miles away:

> *"Once they got started, the cattle moved east quite willingly, and by seven o'clock there were dozens of strings of white-faced, Matador Herefords on a six-mile front, all heading toward the rising sun. Such a glorious sight we shall never see again."*[25]

That was just the start of a month-long drive. On the shores of the South Saskatchewan, the men tried to coax the herd to enter the water at the Saskatchewan Landing ferry site and swim across, but they refused. For three days the Matador men kept trying to get the reluctant Herefords across the river, while the ferryman, Isadore LaPlante, watched from the shore. Isadore, Métis and eighty years old, knew the prairie when it belonged to the bison and those who could hunt them. He had helped drive Matador herds across the river before but he waited before giving advice. Finally, he told the cowhands that cattle do not like to swim in such a large group. Divide them into smaller bunches and they will cross.

As he watched the last herd leave the river and ascend the southern slopes to disappear over the horizon, Isadore may well have thought of the day eighteen years earlier when he guided the Matador's head man, the legendary Murdo McKenzie, across the south branch and east to the hills that would become the company's northernmost holdings. During those years, Hereford steers grazed around the bones of the last wild bison to fatten on their summer range north of the river. According to Graber, the Matador cattle would sometimes stray as far as seventy miles north to where Rosetown is today. Were it not for Ottawa's policies favouring settlement and cultivation from the US border north to the forest fringe, there would have been grass enough for cattle and bison to have their own portions of that vast stretch of land. Isadore and his people, as well as the Cree and Blackfeet consigned to reserves, could have used wild and domestic grazers to manage such large allotments of land across the northern plains.

That possibility, of course, did not fit plans of Manifest Destiny south of the border, nor Canadian dreams of westward expansion, but it took more than colonial schemes to convert the vast prairie world into islands of grass.

All dreams of the unfenced prairie before farmers plowed it up—whether it was a rancher imagining vast open-range for cattle or a Niitsitapi elder invoking the spirit of iinnii—faded after the introduction of a simple but powerful tool of modern agriculture.

People who see bison as a threat and competition look back nostalgically to the romance of open-range grazing, but that era was brought to an abrupt end by the same invention that a generation earlier had sealed the fate of the bison—a tool that ironically has become an icon of modern cattle production. Barbed wire, invented in the

1870s by an Illinois farmer, was the technology that made it possible for farmers to settle and grow unmolested crops on the dry soils of the Great Plains.[26] Without it, you could not protect wheat and barley from large grazers, domestic or wild. With it, you could plunk down a family on every square mile of prairie, turn it grass-side down with a John Deere plow and plant cash crops for export. Once homesteaders started to arrive with that dream in their heads, the land had to be cleared of bison, Indigenous people, open-range cattle, and any other impediments in the way of Western progress.

Yes, the legendary deep snows of 1906-07, destroying more than half the cattle on the northern range of southwestern Saskatchewan and southeastern Alberta, punctuated the open-range era with a full stop, but by then the homesteaders were invading the prairie by the thousands and insisting that all grazing operations be closed in with barbed wire.

Stan Graber finished his story of the last roundup with the now legendary words of James Barnet Henson, a Matador ranch cowhand who had no love for farmers, whom he called "mossbacks". In 1966, Graber says, a boy opened up a tobacco can he found in the nuisance grounds at Lucky Lake, Saskatchewan and found Henson's last will and testament. It was dated September 1919 but amended almost exactly a year after the last roundup, when he added a codicil that reads, "Finally I leave to each & every Mossback my perpetual curse, as some reward to them for their labors in destroying the Open Range, by means of that most pernicious of all implements, the plow. As witness my hand this the 9th day of May 1922."

The long-standing ranching families of the Milk and Frenchman river drainage, the Cypress Hills and the Missouri Coteau see themselves in a lineage with the open-range culture, as the modern counterparts of a prairie way of life that has always been on the margins of the mainstream. They would not have it any other way. Survival from generation to generation is its own reward that is hard-won with independence, tenacity and traditional knowledge that can come only from those who in turn have learned it from their elders.

If they are conservative in their politics and slow to adopt new ways of managing cattle or their pastures, it is only because most of those who made radical changes did not survive to pass on their way of ranching. The narrow ecological and economic margins in which they make a living from the grass do not invite swings

A working cowboy's garb is part cultural identity and part uniform.

of experimentation. Their natural caution has its congener in the prairie itself. In spring, the native forbs and grasses on the northern plains are slow to commit themselves to a great burst of growth. Deep in their cells they know that putting too much growth above ground too early is risky in a climate where one year the skies are stingy with moisture and the next year it snows in June and floods in July.

A hypothetical scenario: ten separate ranchers discover a burrowing owl in their pasture. How do they react? One or two might call Operation Burrowing Owl's "Hoot Line" (1-800-667-HOOT [4668]) to report it. Others decide it is best to keep it quiet, worried that an endangered species on their land will cause them trouble.

They are worried that government people will come and tell them what they can and can't do on their land. Some fear that government agencies and researchers will do more harm than good.

On rare occasions, a rancher might reach for a third option—a final solution.

"Shoot, shovel, and shut up," as rural parlance has it, is not a myth. They are rare, but there are landowners who would sooner kill a burrowing owl than take a chance that someone from a conservation agency might begin to pay attention to the acre of grass containing its nest site, and start offering unwelcome advice. Even if the 3-S solution is just coffee-row bravado, it helps to bleed off some of the frustration and alienation ranchers undergo as they hear mounting concerns for prairie creatures in decline.

Cattle producers are understandably defensive in a time of industry consolidation, declining beef consumption in Canada, and media stories blaming antibiotic resistance and climate change on the cattle industry, without also recognizing the role played by consumer demand for the cheap, marbled product coming out of feedlots.

This side of "shoot, shovel, and shut up" is a whole spectrum of standard defensive talking points I hear from people who raise cattle:

If it weren't for us looking after the grass there wouldn't be any native prairie.... Why would we listen to government people now? It is private ranchers and not governments or conservation groups that have maintained the grassland and saved it from the plow.... I don't need any bureaucrat coming out here to tell me how to manage this land.

Then there are the comments about the species:

Those birds will come back. Everything in nature goes in cycles.... My granddad said there were none of those birds here when he first homesteaded—maybe things are just going back to normal.... It's all those hawk nest platforms they put up—that's what's hurting the birds.... It's all those swift foxes they released.... Endangered species? Hell, I'll show you an endangered species—you're looking at one.

Some of those rangeland proverbs sound a lot better when they are coming from underneath a Stetson or from someone standing on a ridge glistening with

Endangered species like the burrowing owl are sometimes seen by landowners as a liability.

needle and thread grass in midsummer seed. The myth of the fiercely independent rancher who wants nothing more than to be left alone to do what he and the prairie do best—raise cattle on grass—is a compelling one that may be partly true, but the global nature and complexity of today's agricultural economy make it impossible for anyone to be independent of the government and the consumer public.

People who raise livestock on native range are not a type and are as varied as any other subset of the human race, but it is fair to say that certain characteristics and ethics are valued in cattle country. Conservative? Yes. Suspicious of outsiders? Sometimes. But you will not find better people—men and women more humble of spirit, more warm hearted, or more willing to pluck a stranded motorist from the roadside.

A couple of summers ago, a birding friend of mine, Chris Harris, and I drove south to Caledonia-Elmsthorpe Community Pasture. Today that island of grass is being managed by a local group of cattle producers who lease the land from the province, but at the time it was still one of the old federal PFRA pastures. Chris had phoned the pasture manager, Glen Elford, ahead of time to get permission for our visit. He asked Glen about a pair of burrowing owls that had been reported in the pasture that summer. He told me Glen seemed quite knowledgeable and very interested in the birds—not at all unusual for PFRA managers.

"Glen wanted me to know that he wasn't the kind of guy who refuses to report endangered species," Chris said. " 'The more people know about those owls, the better,' he told me."

That evening, we arrived at the pasture's edge along a public road, parked the car and got out. As we stood peering into the expanse of grass heading west and watching chestnut-collared longspurs roller coaster up and down in the air, a black truck slowly pulled up behind us. The door opened our way, showing its Government of Canada logo, and out stepped a large figure in boots, blue jeans, silver belt buckle, and ball cap.

"Chris?" he said, hand outstretched, the other one removing sunglasses as he smiled broadly. Clear-eyed, face bronzed by days in the sun—Caledonia-Elmsthorpe's long-time steward had the bearing, features, and hand-crushing grip I have come to expect of PFRA managers.

Glen told us he was just heading back to pasture HQ where his wife was waiting with supper, but he was in no hurry. We talked about the June rains, the grass, and

he asked what we had seen so far. We pointed to the longspurs, and the soft trills of a Baird's sparrow drifting out across the speargrass.

Eventually, I had to ask the question I ask every PFRA employee: how do you feel about the federal government shutting down the pasture system and turning the land over to the provinces?

"Well, can't say I like it," Glen began, "but there's not a whole lot I can do about it. I've been here a long time. My kids grew up here. When you put your life into a place doing something you believe in, you want to think it will continue after you are done. You don't want to hear that it isn't worth keeping."

When conversation got around, as it inevitably does, to the question of management and the grazing patrons taking control of management decisions, Glen was polite and circumspect, careful not to judge others or speak hastily, but it was clear that he believes that the PFRA quality of management will be hard for most private grazing patron businesses to match.

"It depends," he said, "Sure there are lots of good stewards out there, but there's some bad ones too, people who just don't know better. We get some patrons telling us we should put more cattle on the pasture to use more grass. They mean well, they see all the grass and think it should all be used, but they just don't understand what it takes to keep a pasture healthy from the roots on up."

But, he added, in the province's southwest, where there is a long-standing culture of ranching native grass, there are people with the knowledge to make it work. He is most worried about the pastures like his and others away from the southwest, where the grazing patrons are often mixed farmers who do not have their own native range, nor the experience it takes to manage it.

"Not everyone who owns cattle is a good manager. Some are, some aren't."

Glen got a grounding in good management watching his father ranch on the native grass of their family holdings next to Grasslands National Park, in a region where the low carrying capacity of the grass tends to weed out any poor stewards over time. Adapting that ethic and respect for native range to his work as a PFRA manager, he became known for his interest in birds and other grassland creatures.

"At workshops, sometimes I'd be called on to talk about the burrowing owls we had here [in previous decades they had as many as ten breeding pairs]. I went to a conference on prairie conservation a while back, with lots of people working to figure out how to conserve this kind of land and the wildlife.... Things are not looking good, but I stay positive. There's no point in getting down about it all."

Though supper was waiting, Glen insisted on taking us himself to see the last

Katydid nymph in a western red lily (*Lilium philadelphicum*).

burrowing owl pair at Caledonia-Elmsthorpe. At first, we could see only the one adult, but when I scanned my binoculars across the grass nearby I found five sets of eyes staring back at me. Glen gave a laugh as he looked through Chris's spotting scope at the young owls, freshly out of the burrow. You could hear in his voice a certain proprietary satisfaction in knowing that there would be a good brood of burrowing owls on his pasture this year.

We continued to talk about the owls, about ferruginous hawks, and the nighthawk circling overhead. Just before we parted ways, Glen recalled something he was told long ago: "When the birds start to go, we should pay attention because we will likely be next."

If you and your family have been holding onto native grass and grazing it sustainably for generations, working with biologists and government agencies when they show up with their clipboards and habitat programs; and if you have meanwhile seen others in your region plow their grass under when cattle prices dip and grain prices rise, you might get more than a little upset one day when the government announces a new set of restrictions aimed at the way you manage your pastures.

The way you see it, the rare prairie creatures have only made it into the 21st century because you and other ranchers have stuck with ranching native range even when it wasn't profitable, even when the government was providing incentives to switch to grain production and everyone else was jumping on a tractor and plowing the prairie under.

Val Marie Community Pasture

OPPOSITE: Clustered broomrape (*Orobanche fasciculata*) is a parasitic plant that feeds on the roots of sage species.

Fewer and fewer ranchers tending a smaller amount of native prairie are now left holding the 'species at risk bag'. As the last defenders of native grass, they are naturally going to be upset when they see the government enact laws that focus on what can and cannot be done on the land they graze. After all, it was government policy that brought the plow to the prairie in the first place, and it was government policy that introduced generations of agriculture support programs that continued to cause the cultivation of native grass right through into the 1990s.

That is how the people who live and ranch in ranch country see things and their perspective must be recognized and respected—particularly when government agencies want to partner with private managers to help an endangered species on the edge of extirpation.

In Canada, the greater sage-grouse is gone from British Columbia, and down to somewhere below 150 birds in southeastern Alberta and southwestern Saskatchewan. Estimates say that it now occupies a mere seven percent of its historic range.

A scrap of that seven percent is on rangeland up against the East Block of Grasslands National Park where the cattle belong to one of Saskatchewan's most respected rancher-stewards of native grass, Miles Anderson. A couple of years ago, I met Miles at an indie music concert in the city, where his daughter Kacy was on stage performing with Clayton Linthicum, a young man from a neighbouring ranch family.

We talked a little that night and before we parted ways he mentioned he was heading to a conference.

"I'm off to Salt Lake City next week to talk at a conference sponsored by the Sage Grouse Initiative." He agreed to let me call him about SGI afterwards, to get his impressions of the conference, and talk a bit more about sage grouse and prairie conservation.

A couple of weeks later we finally caught up with one another on the phone and had a good long talk about ranching, stewardship, and the sorry state of sage grouse at the northern edge of their range.

"It might seem like we have a lot of sage grouse habitat here and in the park [Grasslands National Park]," Miles said, "but if you look at their whole range on the continent, we are just a postage stamp. One of the things that makes this land different is that it wasn't glaciated. So it starts right in the soil, and it's part of why we have so many rare species. It's not something I am doing. It's just the way this place is."

Miles and his wife Sheri run cattle on 30,000 acres, including 22 sections of native grass right along the international boundary—all of it excellent habitat for greater sage-grouse and many other species at risk. The sage grouse use a variety of landscapes on his range: wet places such as alkali flats, as well as uplands and lowlands. Miles elaborates, "In winter, they need places where they can feed on sage brush, which means it has to be high enough to poke out of the snow. If the snow's too deep, they move south to find sage brush they can get at."

"My neighbour to the south in Montana was at the conference as well. His land runs from the border almost all the way down the Milk River. He was a guest speaker too and we both had some good airplay."

One of his favourite moments at the conference was when Noreen Walsh, Regional Director with the U.S. Fish and Wildlife Service was speaking. "She said something that agrees with what I see on our land: 'what's good for the birds is good for the herds.'"

"Sure, it's just anecdotal, but I saw some things this past summer that prove the point. It might sound funny but I know what I saw."

The strangest thing he learned that year was that sage grouse seem to like hanging around his cattle. Just over the fence in the East Block of Grasslands National Park the grass had not been grazed by any kind of bovine, wild or tame, in 25 years. Long overdue, the Park had recently welcomed Miles and his Angus cattle onto a portion of the East Block.

The hard part was getting them to stay there once they opened the fence.

"The grass is old and ugly, so we were always having to chase them back into the park." That meant Miles was on horseback and moving cattle during a time of the summer when he does not usually ride among his herd.

A cattle drive through the badlands adjoining the east block of Grasslands National Park.

"We saw sage grouse every day, but we always saw them with the cattle." In his experience, the hens leave the sage brush bottom lands soon after the young hatch, but later in the summer they return with half-grown chicks.

"They might be showing them how to feed on sage—I don't know—and maybe they like the cattle because it helps them find bugs."

Though Miles was excited to see good numbers of sage grouse with young this summer and well into the fall, he knows the next trick is to get good survival into the following year.

Survival is influenced by many factors, but, like many who live in sage grouse country, he is concerned about predators. "We've got ravens like you wouldn't believe. Then the coyotes, hawks, and swift foxes are probably taking their share. And that might be another reason the grouse like cattle around. I've watched them take refuge close to cattle when a harrier flies overhead. You have to ask yourself if there is something we're doing that gives predators an advantage now. Coyotes especially—back in the days when people could control coyotes we used to see hundreds of sage grouse."

But a rancher like Miles Anderson knows there are no simple answers. He thrives by respecting the complex interrelationships between weather, grass, the genetics of his herd, and his own management practices. "One of the speakers at the SGI conference said 'there is no single cause of sage grouse decline. There isn't going to be any single solution.' That seemed right to me."

Toward the end of our call, I asked Miles what he would do if he had a couple million dollars to help out the greater sage-grouse in his corner of Canada's grassland

"That's a tough one, but I guess I would spend it on connecting the science people with ranchers. I think the Sage Grouse Initiative has done some good things in the States. There are a dozen or more programs that a producer might qualify for

down there, but they send out a rep and if you're interested they'll assess your operation and then figure out which programs might work for you. Might be something to do with fencing or infrastructure, might be about water development, deferred grazing, grass banking. All kinds of things. Then they do all the paperwork for you. It's all completely confidential so the ranchers feel safe and they don't have to do all the applying and figuring. The resource person comes back with a program or two and gets it going. I think they get pretty good uptake that way."

Miles added that, while ranchers are naturally cautious and some have had bad experiences in the past with conservation programs, he agrees that building trust and rapport between producers and the conservation and government sectors must be a priority.

"Take the pipit [Sprague's pipit, another species at risk on his grasslands]. You look across the fence and see, well this guy has a lot more pipits than his neighbour. Why not get to know him, spend some time there and learn what he is doing that makes the difference? Then you've got to find a way to encourage more of that, to make it worth other ranchers' time to try something new."

Wild licorice at sunrise.

You don't last in the ranching life as long as Miles has by jumping to conclusions or adopting every new management trend that comes along. Applying the same caution to sage grouse, he is not going to make any rash predictions about its future or claim to know exactly what needs to be done.

And yet, after listening to him discuss the prairie and its declining birds, you go away believing that there might be some hope left. With people like Miles Anderson on the land, and some resources and political will to work with ranchers and industry, is it possible that we could find the mix of private and public management, science-based and traditional stewardship to bring the greater sage-grouse back into its long vacant breeding grounds on the Canadian Plains?

Miles and I never got around to the topic of bison reintroductions. Better to leave that to more skilled hands. The ranchers in the room did not say much the day Paulette Fox described the border-crossing Iinnii Initiative. Miles, who was there at the time, may never become a cheerleader of bison "re-wilding" but he is open-minded in

Sunset at Spy Hill-Ellice PFRA Pasture on the Saskatchewan-Manitoba border.

conversation and tends to keep his powder dry. I have no trouble imagining him and Paulette in a conversation about how to run cattle production and bison restoration side by side in large landscapes crossing international boundaries.

Chief Mountain stands high above the borderland—between alpine and grassland and between Canada and the U.S. Ten years from now or fifty, it will still be there, whenever people find ways to welcome the bison back home.

5

> There are nighthawks here, not the kind
>   that used to ride herd by moonlight, watching
> Herefords that had never seen barbed wire.
>   These ones sleep on the rail
> fence surrounding the homestead.

Possible Prairie

THE YARD SITE where I sit on a porch looking at dozing nighthawks is surrounded by 13,000 acres of prairie that was tended lovingly for decades by a rancher before it became the responsibility of a private conservation non-profit (the Nature Conservancy of Canada [NCC]).

The Peter and Sharon Butala ranch at NCC's Old Man on His Back (OMB) property was mostly Crown land leased and managed privately throughout the 20th century. In 1996, the Butalas were reaching retirement years and became concerned that the land might be plowed to grow crops or degraded by mismanagement or acreage development. That was the year they began the process of passing the land onto NCC by donating the 1,000 acres they held in private title. The remainder of the ranch was subsequently secured in lease acquisitions: public land managed privately for the public good.

The common nighthawks, on Canada's list of species at risk, don't particularly care whether the land is owned or managed privately or publicly. They just like the miles of grass grazed moderately, the bugs that it provides, and the roosts offered by fence lines made of steel pipes to keep the bison away from the yard site. The Sprague's pipits I can hear over the north range are similarly neutral on the question of public versus private.

Ranchers and conservationists are not. Most take a strong and predictable stand on one side or the other. I am here to see what I can learn about that conflict while I look for birds and a bit of grassland conservation history that starts with a man remembering his boyhood on native grass just this side of the U.S. border.

I am joining Ed Rodger, volunteer caretaker for the Important Bird Area that encompasses three vast former Prairie Farm Rehabilitation Administration (PFRA) pastures (Battle Creek, Nashlyn, and Govenlock) which connect to one another in a 940 square kilometer island of native grass that is one of the largest on the northern Great Plains. If all goes well, we will hike into the Battle Creek pasture and find the remains of the homestead where American novelist Wallace Stegner lived as a child. And tonight, when the sun goes down, we will survey nighthawks along a trail that runs through the provincial pasture just west of NCC's OMB ranch.

For many of us who live in the dry western landscapes of this continent, Wallace Stegner is regarded as one of the first to articulate the ecological and cultural destruction in the wake of western frontier-chasing. Saskatchewan people claim a piece of his legacy because he lived for a spell in this province and then, returning for a visit forty years later, wrote *Wolf Willow*, part memoir and part history of the Cypress Hills region.

In 1914, when Wallace was five, his family moved from Iowa to southwest Saskatchewan, spending the six years that comprised the most settled and formative part of Stegner's childhood. His father, George Stegner, was a feckless schemer, always on the move, looking for Eldorado, taking his wife and two sons from frontier town to frontier town, and often leaving them behind for months as he continued to search for a way to get rich fast.

Arriving in southwest Saskatchewan, George took out a homestead quarter and a preemption quarter, 320 acres of prairie wildness just north of the U.S. border. The soil was very poor but they plowed some of it anyway and had a decent crop the first year. In 1916, though, the crop was rained out. For the next three years after that, the grain withered in drought. The Stegners moved back to the States in 1920, abandoning their homestead. Wallace was eleven years old.

Common nighthawks like to roost on fence rails at the Nature Conservancy of Canada's Old Man on His Back Conservation Area.

OPPOSITE: Battle Creek Community Pasture joins with two neighbouring pastures to make up 940 square kms of grassland.

By then, though, he had learned that the prairie earth, with its winds, solitude, dangers, and wild creatures, has a moral and spiritual weight that feeds the soul. As an adult, Stegner's memories of the Saskatchewan prairie would inspire not only some great novels (including the Pulitzer-prize-winning *Angle of Repose*), but help him to launch a Western environmental movement that protected public conservation lands.

In the 1950s Stegner wrote several magazine articles and two books defending public lands from private development and rivers from proposed dams. But his great triumph as a voice for conservation came shortly after returning home from the Saskatchewan trip that forms the heart of *Wolf Willow*.

It happened when Stegner sat down to write a 2400-word letter to the Outdoor Recreation Resources Review Commission in December 1960, never guessing the chord it would strike and the legislation it would inspire. The letter soon fell into the hands of Stuart Udall, the Secretary of the Interior, who used it as the basis of *The Wilderness Act*, a single piece of federal legislation in the United States that continues to this day to defend more than 100 million acres of public lands from any activity or development that would destroy it. If Canada had similar legislation, it may have prevented the federal government from transferring the PFRA pastures back to Saskatchewan where the land is now being managed privately for cattle production without any conservation programming or meaningful forms of protection.

Spring bloom in Missouri Coteau, Saskatchewan.

When the act was finally passed in '64, Stegner's letter was read out in the House of Representatives. At the heart of the letter's eloquent plea is this passage clearly tracing his "idea of wilderness" to his time on the Saskatchewan prairie:

"Let me say something on the subject of the kinds of wilderness worth preserving. Most of those areas contemplated are in the national forests and in high mountain country. For all the usual recreational purposes, the alpine and the forest wildernesses are obviously the most important, both as genetic banks and as beauty spots. But for the spiritual renewal, the recognition of identity, the birth of awe, other kinds will serve every bit as well. Perhaps, because they are less friendly to life, more abstractly nonhuman, they will serve even better. On our Saskatchewan prairie, the nearest neighbor was four miles away, and at night we saw only two lights on all the dark rounding earth. The earth was full of animals—field mice, ground squirrels, weasels, ferrets, badgers, coyotes, burrowing owls, snakes. I knew them as my little brothers, as fellow creatures, and I have never been able to look upon animals in any other way since. The sky in that country came clear down to the ground on every side, and it was full of great weathers, and clouds, and winds, and hawks. I hope I learned something from looking a long way, from looking up, from being much alone. A prairie like that, one big enough to carry the eye clear to the sinking, rounding horizon, can be as lonely and grand and simple in its forms as the sea. It is as good a place as any for the wilderness experience to happen; the vanishing prairie is as worth preserving for the wilderness idea as the alpine forest."[27]

While Stegner succeeded in helping to protect public lands in the United States, the piece of publicly-owned prairie in Saskatchewan that once included his family homestead and inspired him to write the Wilderness Letter is facing an uncertain future. If Battle Creek and the two large adjoining pastures become provincial land as scheduled[28] and transition into private grazing management they will, like all the other federal pastures, lose the conservation programming, and grassland and species at risk management systems that kept the ecosystems in good condition for decades. Private managers in this region will have no difficulty handling the grazing and cattle needs, but they cannot be expected to also manage the complexities and absorb the costs that come with balancing the conflicting ecological needs of more than a dozen species at risk.

Perhaps because of the size of these three pastures and their unequalled density of species at risk, in the summer of 2017, the federal government and Saskatchewan began discussing a land swap deal that could protect Battle Creek, Nashlyn, and Govenlock pastures in a single grassland conservation area. Under the proposal, the land would be managed by Environment and Climate Change Canada (ECCC) in cooperation with the grazing patrons whose livestock use the land. There is no guarantee that the deal will be approved, especially if the cattle producers who graze the pastures are afraid they will lose local control to a federal environmental agency that places conservation above cattle grazing.

A large parcel of federally-managed grassland in one corner of Saskatchewan is better than none, but it will do little for the species and other public values being placed at risk on the remaining two million acres of grassland in one hundred parcels of transitioning federal and provincial community pasture lands in the province. For contrast, just across the border in the western United States, the federal government still spends millions of dollars to protect ecosystems and species and regulate private grazing and other land use on 250 million acres of rangeland. When the oil industry and other business interests begin to lobby for the lands to be transferred to the states, everyone from cattle producers to hunters, hikers and photographers leap to their defence, often using language and arguments first developed by Stegner in his Wilderness Letter.

The sunny face of gaillardia or banket flower (*Gaillardia aristata*).

OPPOSITE: Scarlet mallow (*Sphaeralcea coccinea*)

Battle Creek Community Pasture contains the homestead site that inspired Pulitzer-prize winning author Wallace Stegner's legendary "Wilderness Letter."

People who have been to the Stegner homestead site tell me it is unmarked and unremarkable: some crested wheatgrass and other weeds where his father plowed a strip, a small reservoir on a creek—all of it engulfed by open prairie running across the international boundary. On clear days Montana's Bearpaw Mountains float on haze above the southern horizon.

My first attempt to find the homestead was rained out during a wet summer but our trip this year is hot and dry and the horizon hazy brown with smoke from forest fires blowing in from the Canadian Shield country to the north.

As we pulled the truck into the yard site at the Battle Creek Community Pasture, the manager, Kevin Cherpin, still a federal employee for now, walked out to greet us. He was wincing as he scrolled through the smart phone in his hand.

"Bad news guys. Just got an email from head office. We have to pull all access permits. No one is allowed on any pasture in the southwest until we get some rain. It's just too dry and we don't want a fire. Sorry 'bout that."

"Even on foot? We're walking."

"Nope. Even on foot. Nothing personal. And they told me to tell you that anyone found on the pasture without a permit will be blacklisted for the next season too."

We shifted the conversation, chatted about the changes to come to his pasture if it switches to private management. We said our goodbyes and jumped back into Ed's truck.

As we pulled out of the yard I thought about the famous words that close Stegner's Wilderness Letter—words that are often quoted in defense of public conservation lands:

> *"We simply need that wild country available to us, even if we never do more than drive to its edge and look in. For it can be a means of reassuring ourselves of our sanity as creatures, a part of the geography of hope."*

Hope felt more remote than ever though as we spent the rest of the day driving the public roads on the periphery of the three big pastures. We stopped at intervals to listen and estimate numbers of pipits and longspurs, and we recorded the position of loggerhead shrikes and ferruginous hawks as we travelled, but every part of

Horses, though supplanted by ATVs and trucks for the most part, are still used to work many community pastures and some private ranches.

me wanted to be out of the truck and on the other side of the fence ambling through miles of grass.

Looking at the local mosaic of land tenure and management from the seat of a pickup truck, everything on the prairie appeared to be thriving or suffering at various coordinates within a matrix of private and public interest: privately-owned and managed land, public land managed by government agencies (community pastures), public land leased and managed by private ranchers, privately-owned land managed by a conservation organization (the Nature Conservancy of Canada), and privately-owned land that has been plowed to grow crops.

It is easy to make snap judgements and be wrong when you look at grass and compare one pasture to another. The height of the grass is a poor measure. You might look at a short, close-cropped pasture and think it has been overgrazed compared to another just across the fence, but the cattle might have been moved out yesterday and the range, healthy underground, will soon recover with fresh growth. And a pasture that has not been grazed for a decade will look thick and tall, but the lack of any grazing has reduced its overall diversity of forbs.

As we drove by miles of pasture land, though, the community pastures seemed to have the most grass. Some of the privately managed pastures appeared to be in the same condition, but others showed signs of the land being pushed too hard—patches of bare soil, vegetation grazed down to nothing next to wetlands, and persistent bunches of crested wheatgrass and pasture sage sticking out from billiard table landscapes.

In public, ranchers will often make sweeping statements about stewardship as though all producers are equally good at tending their range, but when you tour the country with them you will sometimes hear a judgement or two: "Oh yeah, he's taking too much there. Not enough carry over." Or, "Not sure who owns that pasture but he's stocking too high." They know that while some ranchers are conservationists, there are some who overgraze and a few who will plow native grass when it suits their needs.

For the most part, though, the men and women who run cattle on leased government land don't expect much from the rest of the world, apart from their purchased lease rights. They prefer to be left alone. If they are to be noticed at all, they might want some recognition for choosing to work with grassland rather than destroy it with a plow. They want conservationists and environmentalists to see the irony and injustice in regulations that penalize ranchers for being the ones who chose a form of agriculture that, when it is done right, protects the oases of prairie remaining in the ecological desert of cropland.

After two long days of travelling around the edges of the pastures south of the Cypress Hills, tired of straining to hear birds on the hot winds of early July, we took a detour to the nearest comfort station I could think of: Manley Bread & Honey Bakery in the village of Consul.

The co-proprietor, David Manley, is also the pastor of the Consul Church of God, but he has taken on the role of community builder in one of the most sparsely populated corners of prairie Canada. David's wife Vicki runs the bakery, and was there when we arrived for tea.

The first time I encountered David, I was in a room full of ranchers fuming over the recently announced Emergency Order for the Protection of the Greater Sage-Grouse (EPO). The Canadian government had done nothing to designate critical habitat for

the species, despite years of study, and the conservation community decided it was time to hold their dragging feet to the fire. Ecojustice launched a court case to force Canada to do something for sage grouse and follow the Species at Risk Act.

When the court decision came down in favour of the grouse, the Conservative government announced measures under the EPO that seemed almost calculated to raise the ire of every rancher in southeastern Alberta and southwestern Saskatchewan. Twenty-six of them were in the Consul community hall that day, glaring across the room at a clutch of fidgeting representatives from the Nature Conservancy of Canada, the Alberta Wilderness Association, Nature Saskatchewan, the U.S. National Fish and Wildlife Foundation, and Environment Canada.

Strong words came out right away as rancher after rancher spoke about the order and how it would hurt their livelihoods and the value of the Crown grassland they lease. They spoke eloquently about their experience with sage grouse and other wildlife, and said that they felt betrayed by the conservation community because the EPO seemed to attack the very people who are living on the prairie and practicing a form of agriculture that prevents it from being plowed and seeded to crops.

There was little any of us could say to relieve the tension in the room. But then another man spoke and that seemed to help. It was David Manley. He was not a rancher but the local pastor, he said, and his manner and words moved the discussion away from the narrow binary of ranchers versus sage grouse to embrace the wider concerns of community.

Later that year I sat down to enjoy a cinnamon bun and conversation with David in the tea room of the bakery where he and his wife are building community with good food and a place for people to sit and visit.

"We've been here for one hundred years. We stayed. I told them that when we met with Environment Canada: the land chose us, maybe it chose us," he said, looking me in the eye to see if such a bold statement might elicit a response.

"We have one hundred years of social capital here. You can't replace that."

His education (Doctor of Ministry Degree in Rural Development) and experience as a pastor in ranch country have helped him to develop an analysis of the gap between ranchers and conservation organizations. In the middle of a conversation

about market forces and other external pressures that are making it harder for families to continue ranching on native grassland, I asked him to explain a mystery to me: almost all ranchers in the area say they would never plow the prairie and hate to see others do it, but most of them don't like the Nature Conservancy of Canada purchasing land, and they say conservation easements are too restrictive.

He responded by comparing the buying of grassland and easements for conservation in the area to the medieval practice of purchasing indulgences to pay for sins—implying that NCC is helping urban people expiate their ecological sins, but doing little to support a culture of conservation in grassland regions.

"Instead, why not invest in community development that will encourage stewardship of grassland?" Like many people in remote regions where there are concerns over the environment, David said he too often sees the experts arrive with a study that is designed to meet outside agendas. They get access to land, talk to landowners and managers, study the wildlife, and then leave without the results benefiting the community.

"We need scientists who are from here, who live here and do research that will benefit those of us who live here."

That is a compelling view from the grassroots local level that needs to be taken seriously. But here is something from the global end of the spectrum, which needs to be heard as well.

In 2008, 35 grasslands experts from 14 countries gathered in Mongolia for the World Temperate Grasslands Conservation Initiative, a subcommittee of the World Council on Protected Areas. They signed an agreement called the Hohhot Declaration, part of which reads as follows:

> *"We the participants of the Hohhot World Temperate Grasslands Conservation Initiative Workshop from five continents and 14 countries, declare that temperate indigenous grasslands are critically endangered and urgent action is required to protect and maintain the services they provide to sustain human life. We call upon all sectors of society to collaborate towards this goal."*[29]

The Auvergne-Wise Creek Community Pasture, one of the federal pastures the Canadian government has handed over to Saskatchewan, hosts several species at risk.

Bill Henwood, one of the Canadians participating in the workshop in Mongolia, put down his thoughts on why he believes the declaration is needed:

> *"After cradling the needs of humans for centuries, temperate grasslands are now considered the most altered and endangered ecosystem on the planet. For most of the past century, temperate grasslands have not been visible on the global conservation agenda. The grasslands used to be home to some of the greatest assemblages of wildlife the earth has ever witnessed. Potential for protection still remains, especially in the prairies of North America, the pampas of South America, the lowland grasslands of southeast Australia, the steppes of Eastern Europe, and the Daurian steppes of East Asia."*[30]

The Hohhot gathering found that less than five percent of the planet's indigenous grasslands receive any form of protection from the forces that continue to threaten them. Since then, that figure has been reduced in another study to 3.9 percent.[31] In Saskatchewan, among the worst on the planet for grassland protection, that figure drops to 2.6 percent[32] once you remove the former federal community pastures. A 2017 budget decision to close the province's own Saskatchewan Pastures Program may well lead to the sale of several hundred thousand acres more, further eroding the protection of Canada's prairie ecozone.

Ranchers may do their best to protect the grassland by grazing it carefully in a form of agriculture that justifies its preservation. But that form of protection over time is not perfect. If it were, we would not be down to less than 20 percent of our native prairie in many regions. Some ranchers do not like talk of protection because they feel it is directed at them and their cattle. But the protection that grassland experts and conservationists advocate is not about protecting land from grazing. It is about protecting the land *and its stewards* from the market forces and agricultural policy that push grassland to more profitable kinds of exploitation that will destroy or degrade the ecology.

Here is an excerpt from Stegner's Wilderness letter again—this time on grazing as part of the wilderness ethic:

> *"... as for grazing, if it is strictly controlled so that it does not destroy the ground cover, damage the ecology, or compete with the wildlife it is in itself nothing that need conflict with the wilderness feeling or the validity of the wilderness experience. I have known enough range cattle to recognize them as wild animals; and the people who herd them have, in the wilderness context, the dignity of rareness; they belong on the frontier, moreover, and have a look of rightness. The invasion they make on the virgin country is a sort of invasion that is as old as Neolithic man, and they can, in moderation, even emphasize a man's feeling of belonging to the natural world."*

I left the Bread & Honey Bakery that day, wondering if we will ever bridge the divide between ranchers and conservationists on the question of how best to protect our remaining native grassland.

Grasslands National Park cannot on its own be expected to represent all the grassland eco-types that are losing their biodiversity and integrity.

Land securement through purchase and easements is admittedly a piecemeal, stopgap effort. No one in the conservation community believes it is enough on its own. Our endangered grasslands and the creatures that depend on them need solutions that will work across large landscapes in a mosaic of public and private land that is for the most part managed by private ranching families. While a majority of the surviving native grassland in the prairie region encompassing western North Dakota, Montana, southwestern Saskatchewan and southeastern Alberta is on publicly-owned land, the day-to-day decisions of how much to graze, how to manage fencing, grass and water are made by private ranchers who live there and use that land.

When conservation people talk about protection, it is not to protect the land from private grazing, but to protect these endangered ecosystems, and the ranching culture that can sustain them, from the many forces that are degrading and destroying more native grassland every year: the economic pressure to subdivide and sell ranchettes, commodity prices and new crop varieties that make it profitable to plow more native grassland, market forces and government lease terms that cause ranchers to overstock, and poorly regulated and monitored industrial development, whether it is oil and gas, gravel or wind energy.

Private cattle producers in true ranch country have been applying their own model of protection based on a culture and tradition of stewardship passed down from one generation to the next—with little public support and in the face of government policy that subsidized the plow. This vital element of native grassland protection is what carried much of our remaining large pieces of mixed-grass and moist mixedgrass prairie into the 21st century.

However—and this is where conservationists and ranchers often part ways—that very ethic of private stewardship seems to protect the prairie best when the land is either utterly unsuitable for cultivation or under public title.

The beauty of keeping grassland in the public domain and leasing it out to private cattle producers is that it leaves room for public policy to help with the handoff of land stewardship from one producer's lifetime to his successor while ensuring that over time the land will not be significantly altered.

The chink in the armour of entrusting grassland conservation entirely to the culture of private rancher stewardship is that even the best steward will die some

Some winters provide only a shallow layer of snow on the prairie. Grasslands National Park, Saskatchewan.

day and he or she may not have an apprenticing relative to take over the legacy. Public ownership or even ownership by a non-governmental organization, such as the Nature Conservancy of Canada, allows us to retain an element of public involvement that keeps the native grass right side up regardless of how heavily or lightly the next leaseholder chooses to graze.

In that way, the public can play a vital role in ensuring that our last remnants of native grass do not become ranchettes or cultivated fields. The quality of our remaining large tracts of native prairie is a matter of how we strike the bargain between

Prairie people need to meet out on the land and talk about ways to balance public and private interests in grassland.

public and private interest on state or provincial-owned grasslands. The individual cattle producers need social licence and affordable lease rates and programs that help them sustain the ecological goods and services produced on rangeland, while the public and conservation groups need grazing animals and their managers to provide the kind of disturbance essential to both healthy grassland and livestock rearing. This is the basis for grassland conservation that is working in various forms and to varying degrees of success all over the prairie landscapes of North America—on community pastures, National Grasslands, provincial and national parks, leased public lands, and public grazing reserves: our remaining grassland commons.

The history of grazing domestic livestock on this continent began with people using the commons without using it up. In that short spell of time between the signing of the treaties and the rush of homesteading farmers, the water, soil, and grass belonged to everyone and to no one, but it could be grazed in private leases at costs proportionate to limited terms of use that recognized the public's common interest in the land's long-term well-being. That ended when widespread settlement and landbreaking by farmers forced a kind of enclosure of the prairie commons that unfortunately is being intensified today by policy-makers ideologically opposed to public land and land management for the common good.

The sustainable relationship between private and public interest suffers every time a government sells more public land, privatizes management of a community pasture, or ignores a farmer's encroachment on public road allowances and other rights of way.

There is great potential to restore the partnership between private and public interest on our native grasslands. As long as they remain in the public domain, community pastures are ideal testing grounds for a new dispensation that would restore the faith and trust between all parties and provide internationally legitimate protection for several thousand square kilometres of grassland. It would take some leadership from government (Canada's federal government in particular owes our grasslands at least that much) to bring all parties to the table and hammer out an agreement or create a pilot project that would be scalable for application on any piece of grassland leased for grazing.

Community pastures, in particular, are landscapes where many of our common values as prairie people can intersect across binaries that too often divide us: rural and urban, producer and consumer, Indigenous and non-Indigenous, private and public. With the right vision and governance, these lands, managed for the common good, could diversify and strengthen farm and ranching communities, advance models of agro-ecology, improve carbon sequestration and climate change adaptation, protect biodiversity and endangered species, welcome medicine-gathering and hunting by First Nations and Métis people, and provide hunting and recreation for non-Indigenous people. Anchored by the traditional knowledge in our ranching and Indigenous communities, and further informed by the best that ecological land management science has to offer, our community pastures could one day be regarded as the seeds of a prairie culture that turned away from extraction and toward sustainability; that renewed its engagement with the commons and found ways to care for land and waters as legacies inherited and held in trust for the well-being of prairie dwellers yet to come.

On the final night of our camp-out at Old Man on His Back, I sat on the grass in front of the headquarters, listening to Ed play guitar from the porch. It had been another hot day, with forest fire smoke making it somehow even hotter. We had found young ferruginous hawks almost ready to fledge from the nest, a family of loggerhead shrikes, a long-billed curlew, and finally, towards dusk, on a nearby pasture, a single burrowing owl.

And, as the sun drew nearer the horizon, the nighthawks left their daytime roosts one by one and flew out over the grass to hunt.

Ed is an accomplished classical guitarist and the song he was playing was heartbreakingly somber, a Latin American melody that tugged at me with its softer notes and pauses. I asked him what it was and he said a friend had composed it to go with some lines of verse written by a woman imprisoned in Chile during the Pinochet era.

He knows Spanish and loves to travel to the Southern Cone grasslands of Argentina, Uruguay and Paraguay, where cattle producers work closely with the conservation community to protect their native grasslands. He paused to translate a line of the verse.

"I don't know all the words that go with the song, but it ends with 'then I will be a wanderer in virgin grasslands.'"

The islands of grass all around us still remember their original wanderers. Deep in our own DNA, and in our dreams, even in those of us whose ancestors came across oceans to this land, there stirs a memory of the virgin grasslands that made us all human. The prairie eye has not left us and was there on all sides when our forebears stood on these plains and signed sacred treaties, promising to share and care for the gifts of the land.

Until the late nineteenth century, plains bison and the Indigenous people who hunted them sustained the health of the continent's grasslands.

We are coming to understand that the promise of the treaties has not yet been fulfilled. And with that understanding there is at least the possibility that we might find ways to go forward and feed ourselves from the bounty of this land without destroying the birds and other creatures who had co-existed with bison hunting peoples for millennia, and who have been suffering ever since we took exclusive title in abrogation of those treaties. What would it take for us to begin to live in the spirit of the treaties, with reverence and respect guiding our every relationship, nation to nation, and people to land?

Miracles? Perhaps, but nature specializes in miracles when we let it lead the way. In this part of the world, that means taking our lead from the grass that has always known how to feed the land and its dwellers.

Appendix: A Call to Action

If you want to help, here are ten practical things you can do:

1. First, and perhaps most important, go experience the native prairie and the people who live and work there. Even better, bring a friend who has never walked through a big stretch of wild grass.

    Summer is a good time to head out and find an expansive piece of native grassland where you can go for a long walk. If you live in places like Alberta or Montana or North Dakota it is fairly easy. There are established trails on state and federal grasslands where you are welcome to hike, picnic, and in some cases camp overnight.

    In Saskatchewan, however, there are very few places where you are allowed to hike and camp on our publicly-owned native grassland. That has to change and should be a priority for grassland advocates. If we want to see more native prairie in the province protected, more people will need to become inspired by first-hand experience of the grassland world.

    On leased Crown grassland or community pastures in the province, you have to track down the leaseholder or manager to ask for permission. In most cases, they will grant the permission and let you know where you can go and where you cannot go, and how to behave while you are there.

    As for the former Prairie Farm Rehabilitation Administration pastures now passing into the hands of private groups of cattle producers who are leasing the land for grazing, there is no formal process for gaining access and it remains to be seen how each pasture corporation and their managers handle public requests for access for recreation, research or traditional medicine gathering.

Fortunately, and thanks to the foresight of Nature Saskatchewan and the great and wise George Ledingham, Saskatchewan does have one big and beautiful piece of publicly-owned grassland where you don't need to ask permission to go for a walk—Grasslands National Park (GNP).

Most GNP visitors restrict themselves to the shorter trails in the West Block, which have their appeal but cannot compare to the grandeur of the landscapes you will encounter in the East Block or the new Gillespie range added to the east side of the West Block. One of the longer and most rewarding East Block hikes is the Butte Creek/Red Buttes Trail, which departs from the Rock Creek Campground and day use area.

The total distance of the hike from Rock Creek Campground is 16 kilometres round trip. As you move through a mixture of habitats, from native grasslands to creek valleys to badlands, there will be birds and a mix of native grasses and wildflowers along the way. You could well see a long-billed curlew, or a prairie falcon pass overhead; ferruginous hawks and golden eagles nest in the area. If you go in June or July the characteristic prairie songbirds will sing all around you and overhead: in open grasslands you will hear Sprague's pipits, chestnut-collared longspurs, Baird's sparrows, and grasshopper sparrows. As you move through the badlands, you may hear the rock wren's improvised song echoing from butte to butte.

And when the day is over, you may know something of what Stegner was talking about: that we need such wild country, if nothing more than to restore ourselves with the reassurance that comes from a "geography of hope."

2. Send a message with the food you eat. A couple of years ago, I asked Canadian Wildlife Service biologist Brenda Dale what consumers can do to help grassland birds and her answer was simple: buy grassfed beef.

"If even 10 percent of people made the change to grassfed beef it would make a big difference. There would have to become more grass out there in order to raise beef entirely on grass and that would be a change driven by consumers that could greatly benefit grassland birds."

Grass-fed beef, however, is not on the menu at most restaurants and you won't find it at most major grocery chains. Virtually all of the meat and dairy products available at our supermarkets come from animals that have been fed a lot of grain. Even animals that graze in native grassland for their first year or two of life usually end up being "finished" with several months of intensive grain feeding at a large feedlot. When it comes to ecological costs, nothing can pile it deeper and higher than a large industrial feedlot. Ranchers, policy-makers, political leaders and consumers have an opportunity to put grassfed beef on the market and find new common ground between economics and ecology in grassland.

Take the time to find out how the agricultural methods producing the food you eat affect grassland species such as migratory birds, butterflies and pollinators. Use your consumer voice and dollars to purchase food grown sustainably and to encourage food growing and gathering practices that nourish the environment.

3. Include First Nations, tribal organizations, and Métis people, as well as cattle producers, government agencies, conservation groups, consumers and city folk in all discussions of how publicly-owned grasslands should be conserved, accessed, shared and used. This is a simple thing, but is often left aside when people look for ways to protect the prairie.

4. Donate to prairie conservation. Ninety-seven percent of charitable giving goes to human causes. Of the remaining three percent, half goes to pets. That leaves one and a half percent devoted to the rest of nature. There are several non-profits in North America doing good work in grassland conservation and restoration—they all need more support.

5. Join an organization that advocates for grassland conservation. Nature Saskatchewan, Public Pastures—Public Interest, the Alberta Wilderness Association, the Nature Conservancy of Canada, your local

wildlife federation and its provincial and national affiliates, Ducks Unlimited Canada, the National Audubon Society, Nature Canada, the Canadian Parks and Wilderness Society, the World Wildlife Fund, the Wilderness Society, the Wildlife Conservation Society, the Grasslands Conservation Council of British Columbia—these are merely some of the many fine non-governmental organizations that work hard to advance the conservation of native grassland ecosystems, by recognizing the positive efforts of ranching families and advocating to maintain public grasslands as publicly-owned and maintained as healthy ecosystems for future generations.

6. If you own a parcel of native grassland or know someone who does, you can help. Whether it is 40 acres or four thousand you can look for ways to support positive management practices. Strive for a patchy diversity of grazed and lightly grazed areas, maintained by keeping stocking rates low. Welcome visitors, especially Indigenous people, and anyone who will walk respectfully on the land without disturbing livestock. Control invasive species. Water livestock away from riparian areas. Control vehicle access, especially ATVs (all-terrain vehicles). When it comes time to pass ownership on, consider a conservation easement to protect the land from sub-division development or cultivation after you sell.

7. Work with your school and school board. Ask them to incorporate grassland education into the curriculum and school activities.

8. Call on your political representatives to spend some money on grassland conservation programming. If we want to protect rare grassland habitats and species without eroding the income of producers who agree to apply certain management practices, our state, provincial and federal governments will have to commit some significant funding. In the last ten years, the state of Wyoming has spent nearly eight million dollars working with ranchers and the resource industry to implement conservation programs aimed at keeping the greater

sage-grouse off the endangered list. Over that same period, Saskatchewan, with twice the human population and government budget, has spent almost nothing to conserve or restore greater sage-grouse habitat even as the bird moves rapidly from severe endangerment to extirpation.

9. Work with local urban and rural municipal governments. They have a role in conserving local wild spaces, creating environmental reserves, advocating for grasslands and pastures that serve your community, and promoting grasslands and other natural areas as assets for your community.

10. Encourage prairie restoration. Once at Grasslands National Park I saw a machine with a rotating drum on the back bristling with thousands of speargrass seeds. They use it to collect native grass seed for restoration projects on park land that had been cultivated. In the last decade, Parks Canada and its corps of local volunteers have seeded 630 acres back to native grass. NCC has done some good restoration work at the OMB property. Any hope of helping grasslands ecology recover some of its former glory will require large scale restoration of prairie ecotypes in a patchy diversity. No one has the science to restore grassland to the fully functioning regime that greeted settlers when they first put it to the plow, but the more researchers and park managers try to restore native grassland the more they learn about how to create a facsimile of the original.

Notes

Chapter 1: The Prairie Eye

1. Bill Holm, *The Music of Failure* (Minneapolis, Minnesota: Prairie Grass Press, 1990).

2. Gregory J. Retallack, "Global Cooling by Grassland Soils of the Geological Past and Near Future," *Annual Review of Earth and Planetary Sciences* 41 (2013): 69–86.

3. Iain J. Gordon and Herbert H.T. Prins, eds., "The Ecology of Browsing and Grazing," (New York: *Springer Science + Business Media*, Sept. 14, 2007).

4. William R. Reynolds, "Energetics and the Evolution of Brain Size in Early Homo," in *Guts and Brains: An Integrative Approach to the Hominin Record*, ed. Will Roebroeks (Amsterdam: Amsterdam University Press, 2007), 36.

5. T. Edward Nickens, "Vanishing Voices," *National Wildlife*, (Sept. 15, 2010), https://www.nwf.org/News-and-Magazines/National-Wildlife/Birds/Archives/2010/Grasslands-Birds-Disappearing.aspx.

6   Bob Peart, *Life in a Working Landscape: Towards a Conservation Strategy for the World's Temperate Grasslands, A Record of The World Temperate Grasslands Conservation Initiative Workshop, Hohhot, China—June 28 & 29, 2008* (Vancouver, British Columbia: International Union for Conservation of Nature, Temperate Grasslands Conservation Initiative, August 2008), https://www.iucn.org/sites/dev/files/import/downloads/grasslandsbrochureapril09.pdf.

7   C. Jenkins and L. Joppa, "Expansion of the global terrestrial protected area system," *Biological Conservation* 142 (10) (2009): 2166–2174, doi:10.1016/j.biocon.2009.04.016.

8   S.M. Evans, *The Bar U and Canadian Ranching History* (Calgary, Alberta: University of Calgary Press, 2004), 38.

9   United Nations, *United Nations Declaration on the Rights of Indigenous Peoples*, (New York, March 2008), Article 27, http://www.un.org/esa/socdev/unpfii/documents/DRIPS_en.pdf.

10 United Nations, *United Nations Convention on Biological Diversity, Aichi Biodiversity Targets*, 2010, Target 18, https://www.cbd.int/sp/targets/default.shtml.

11 G. Riemer, "Agricultural Policy Impacts on Rangeland and Options for Reform: An Overview and Evaluation," in *Proceedings of the First Interprovincial Range Conference in Western Canada. Managing Canadian Rangelands for Sustainability and Profitability.* (Regina, Saskatchewan: Grazing and Pasture Technology Program, Saskatchewan Stockgrowers Association, 1994).

12 Chris Nykoluk, "What are native prairie grasslands worth? Why it Pays to Conserve this Endangered Ecosystem," (Saskatchewan: Ranchers Stewardship Alliance Inc., 2013).

13 Jennifer Paige, "Study shows grassland environmental contributions," *Manitoba Co-operator*, (December 7, 2016), https://www.manitobacooperator.ca/livestock/study-shows-the-extent-of-grassland-environmental-contributions.

14 Stephenie Ambrose Tubbs and Clay Jenkinson, *The Lewis and Clark Companion: An Encyclopedic Guide to the Voyage of Discovery*. (New York: Henry Holt & Co., 2003).

15 Lawrence J. Barkwell, "Frenchman Creek Metis Wintering Camp," (Louis Riel Institute), from Philip Rappaglosi and Robert Bigart, *Letters from the Rocky Mountain Indian Missions: Philip Rappaglosi* (Lincoln, Nebraska: University of Nebraska Press, 2003): xxxvii-xxxviii, https://www.scribd.com/document/202804118/Frenchman-Creek-Metis-Wintering-Camp.

16 George Ledingham, "Prairie Dogs and Ferrets," *The Blue Jay* (March, 1965): 2 (inside cover).

17 Steve Zack and Kevin Ellison, *Grassland Birds and the Ecological Recovery of Bison: A Conservation Opportunity* (Bozeman, Montana: The Wildlife Conservation Society in association with the American Bison Society), n.d. http://www.eco-index.org/search/pdfs/1354report_1.pdf.

18 Nickens, op. cit.

19 Henry Youle Hind, *Narrative of the Canadian Red River exploring expedition of 1857 and of the Assiniboine and Saskatchewan exploring expedition of 1858* (London, England: Longman, Green, Longman and Roberts, 1860).

20 Dale Hjertaas, "Summer and Breeding Records of the Whooping Crane in Saskatchewan," *The Blue Jay* (June 1994) 52(2): 99–115.

21 Saskatchewan Department of Agriculture, *Annual Report of the Department of Agriculture of the Province of Saskatchewan*, 1922 (Regina, Saskatchewan), 291.

22 Arthur Cleveland Bent, *Life Histories of North American Marsh Birds* (Washington: Government Printing Office, 1926), 222–24.

23 Ipsos Reid Poll for the Canadian Wildlife Federation, Dec. 12, 2012.

24 It needs to be said, however, that the coexistence of cattle and bison had its darker side as well. As James Daschuk has pointed out in *Clearing the Plains: Disease, Politics of Starvation, and the Loss of Aboriginal Life* (Regina, Saskatchewan: U of R Press, 2013), cattle brought bovine tuberculosis to the plains, infecting the wild bison herds and Indigenous people who received government rations of tainted beef.

Chapter 4: Islanders: People of the Grassland

25 Stan Graber, *The Last Roundup* (Saskatoon, Saskatchewan: Fifth House, 1995), 45.

26 Chris Magoc, *Imperialism and Expansionism in American History: A Social, Political, and Cultural Encyclopedia and Document Collection* [four volumes] (Santa Barbara, California: December 2015, ABC-CLIO).

Chapter 5: Possible Prairie

27 Wallace Stegner, "Wilderness Letter," (Los Altos, California: December 3, 1960), http://www.colorado.edu/AmStudies/lewis/west/wilderletter.pdf. Also, written to the Outdoor Recreation Resources Review Commission, and subsequently in his "Wilderness Idea," in *The Sound of Mountain Water* (Garden City, N.Y.: Doubleday, 1969.)

28 Half of Govenlock pasture, however, is federal land and not scheduled to revert to the province.

29 Peart, op. cit.

30 Bill Henwood, "Towards a Conservation Strategy for the World's Temperate Grasslands," (Vancouver, British Columbia: World Commission on Protected Areas and the International Union for Conservation of Nature, Temperate Grasslands Conservation Initiative, 2010), https://www.iucn.org/sites/dev/files/import/downloads/tgci_strategy_paper_2010.pdf.

31 C. Jenkins and L. Joppa, "Expansion of the global terrestrial protected area system," *Biological Conservation* (2009) 142(10). 2166–2174, doi.10.1016/j.biocon.2009.04.016.

32 Seth Fore, Kate Overmoe and Michael J. Hill, "Grassland conservation in North Dakota and Saskatchewan: contrasts and similarities in protected areas and their management," *Journal of Land Use Science* (2015) 10:3: 298–322. Published online Dec. 2013, doi: 10.1080/1747423X.2013.858787.

Acknowledgements

First, we would like to acknowledge the many people—some of whom appear in this book—who have informed our travels through the islands of grass remaining on the northern Great Plains over the last few years. The ranchers, biologists and conservationists who think about native grassland may not always agree on solutions, but there is a piece of common ground—their passion for keeping the prairie grass-side up—that may yet one day help these islands of grass to flourish and even expand.

A special word of thanks to the photographers who allowed us to use particular images of plants and animals.

We are particularly grateful to our editors. Bruce Rice and Joanne Havelock brought so many fresh thoughts and insights to the manuscript and have improved it immeasurably. Their firm grasp on the nuances and mechanics of writing would have been enough, but they both understand the complexities and issues around grassland, and that knowledge has enhanced the narrative on nearly every page. Thanks also to Joyce Clark for her excellent copy edit. And finally we would also like to express our gratitude to John Agnew, Susan Buck, Tania Craan, Mackenzie Hamon and the rest of the team at Coteau Books. We would especially like to thank Tania Craan for her wonderful design—and for her patience with us during the editing process. We believe they have all helped us put together a book that honours our native prairie, its wild creatures, and the people who are looking for ways to balance private grazing interests with the long term public interest in conservation.

Photographic Information

All images in this book are by Branimir Gjetvaj unless otherwise indicated.

Table of Contents – *Sunset in Grasslands National Park*. Val Marie, Saskatchewan. Note: mirror image of the original.
Page 1 – *Horses in Pasture*. Zortman, Montana.
Page 2 – *Blanket Flower in Prairie by South Saskatchewan River*. Near Empress, Alberta.
Page 3 – *Swift Fox* by Nick Saunders.
Pages 4-5 – *Frenchman River Valley*. Grasslands National Park. Val Marie, Saskatchewan.
Page 7 – *Prairie at Old Man on His Back Prairie and Heritage Conservation Area*. Claydon, Saskatchewan.
Page 8 – *Sunset at Old Man on His Back Prairie and Heritage Conservation Area*. Claydon, Saskatchewan.
Page 9 – *Horses in Auvergne*. Wise Creek Community Pasture. Cadillac, Saskatchewan.
Page 10 – *Prairie with Sagebrush*. Zortman, Montana.
Page 12 – *Coyote in Winter*. Grasslands National Park, Val Marie, Saskatchewan.
Page 13 – *Aspens in Winter*. Wolverine Community Pasture. Plunkett, Saskatchewan.
Page 14 – *Organic Farm bordering Reno Community Pasture*. Robsart, Saskatchewan.
Pages 16-17 – *Farm Fields near Biggar*. Biggar, Saskatchewan.
Page 18 – *Native Prairie at Masefield Community Pasture*. Monchy, Saskatchewan.

Page 21 – *Aerial View of South Saskatchewan River Valley near Leader*. Leader, Saskatchewan.
Page 24 – *Winter in the East Block of Grasslands National Park*. Val Marie, Saskatchewan.
Page 26 – *Sunset at Waldron Ranch*. North of Lundbreck, Alberta.
Page 29 – *Native Prairie at Masefield Community Pasture*. Monchy, Saskatchewan.
Page 30 – *Herding Cattle on Anderson Ranch*. Fir Mountain, Saskatchewan.
Page 33 – *Abandoned Farm Machinery*. South of Abbey, Saskatchewan.
Page 34 – *Native Prairie with Buffalo Bean*. Old Man on His Back Prairie and Heritage Conservation Area. Claydon, Saskatchewan.
Pages 36-37 – *Native Prairie at Fairview Community Pasture*. South of Fiske, Saskatchewan.
Pages 38-39 – *Oil Wells on Swift Current - Webb Community Pasture*. Swift Current, Saskatchewan.
Page 39 – *Cattle Sorting at Wolverine Community Pasture*. Plunkett, Saskatchewan.
Page 40 – *Sunset at Grasslands National Park*. Val Marie, Saskatchewan.
Pages 42-43 – *Abandoned Car and Storm Clouds*, Caledonia-Elmsthorpe Community Pasture. Avonlea, Saskatchewan.
Page 45 – *Old Man on His Back Prairie and Heritage Conservation Area*. Claydon, Saskatchewan.
Page 47 – *Wetland Surrounded by Native Prairie*. Cypress Hills near Robsart, Saskatchewan.
Page 49 – *Marsh at Shallow Lake*, Progress Community Pasture. Luseland, Saskatchewan.

Page 54 – *Cushion Milk-vetch* by Glen Lee.
Page 60 – *Dung Beetle* by Janet Ng.
Page 61 – *Long-billed Curlew* by Annie McLeod.
Page 62 left – *Mormon Metalmark* by Johane Janelle.
Page 62 right – *Mormon Metalmark* by Sherri Grant.
Page 64 left – *Baird's Sparrow* by Nick Saunders.
Page 64 right – *Chestnut-collared Longspur* by Nick Saunders.
Page 66 – *Yellow-bellied Racer* by Janet Ng.
Page 67 – *Whooping Cranes* by Annie McLeod.
Page 69 – *Whooping Crane* by James Villeneuve.
Page 75 – *Burrowing Owl* by Nick Saunders.
Page 77 – *Tipi in Morning Fog*, Wanuskewin Heritage Park. Saskatoon, Saskatchewan.
Page 78 – *Tipi Rings in Grasslands National Park*. Val Marie, Saskatchewan.
Page 80 – *Herd of Bison in Grasslands National Park*. Val Marie, Saskatchewan.
Page 81 – *Moving Cattle*. South of Fir Mountain, Saskatchewan.
Page 82 – *Tire Tracks in Prairie at Sunrise*. Cypress Hills, Saskatchewan.
Page 85 – *Moving Cattle Between Pastures*. South of Glentworth, Saskatchewan.
Page 86 – *Herding Sheep on Waldron Ranch*. North of Lundbreck, Alberta.
Page 87 – *Cattle at Govenlock Community Pasture*. Govenlock, Saskatchewan.
Page 89 – *Cattle Chute and Big Dipper*. Great Sand Hills south of Abbey, Saskatchewan.
Page 90 – *Corrals in Grasslands National Park*. Val Marie, Saskatchewan.
Page 92 – *Large boulders in Prairie*. Grasslands National Park. Val Marie, Saskatchewan.
Page 95 – *Rain Storm Passes through Grasslands National Park*. Val Marie, Saskatchewan.
Page 97 – *Prairie with Sage Brush in Val Marie Community Pasture*. Val Marie, Saskatchewan.
Page 98 – *Val Marie Community Pasture*. Val Marie, Saskatchewan.
Page 100 – *Cattle Drive near East Block of Grasslands National Park*. Glentworth, Saskatchewan.
Page 103 – *Wild Licorice at Sunrise*. Great Sand Hills near Piapot, Saskatchewan.

Page 104 – *Cowboy Boot Fence*. Great Sand Hills south of Sceptre, Saskatchewan.
Page 105 – *Sunset at Spy Hill-Ellice Community Pasture*. Binscarth, Manitoba.
Page 106 – *Twilight at Old Man on His Back Prairie and Heritage Conservation Area*. Claydon, Saskatchewan.
Page 108 – *Native Prairie at Battle Creek Community Pasture*. Divide, Saskatchewan.
Pages 110-111 – *Wildflower Bloom in Missouri Coteau*. Spring Valley, Saskatchewan.
Pages 114-115 – *Battle Creek Community Pasture*. Divide, Saskatchewan.
Page 118 – *Prairie at the Edge of Farm Field*. South of Empress, Alberta.
Page 121 – *Sunset at 70 Mile Butte*. Grasslands National Park. Val Marie, Saskatchewan.
Pages 122-123 – *Morning Mist in Grasslands National Park*. (2 views) Val Marie, Saskatchewan.
Page 125 – *Native Prairie at Auvergne-Wise Creek Community Pasture*. Cadillac, Saskatchewan.
Page 126 – *Prairie with Wildflowers*. East block of Grasslands National Park, Saskatchewan.
Page 128 – *Winter in Grasslands National Park*. Val Marie, Saskatchewan.
Page 129 – *Winter in Grasslands National Park*. Val Marie, Saskatchewan.
Page 130 – *Visitors to Grasslands National Park*. Val Marie, Saskatchewan.
Page 133 – *Hummocky Moraine at Old Man on His Back Prairie and Heritage Conservation Area*. Claydon, Saskatchewan.
Page 134 – *Bison in Grassland National Park*. Val Marie, Saskatchewan.
Page 135 – *Sunset at Old Man on His Back Prairie and Heritage Conservation Area*. Claydon, Saskatchewan.
Page 141 – *Native Prairie at Masefield Community Pasture*. Monchy, Saskatchewan.
Page 146 – *Stallion Running through Prairie in Grassland National Park*. Val Marie, Saskatchewan.
Page 148 – *Sunset over South Saskatchewan River*. South of Empress, Alberta.

About the Authors

Trevor Herriot is a prairie naturalist, activist and writer living on the northern edge of the Great Plains in Regina, Saskatchewan. He has five previous titles to his name, most recently *Towards a Prairie Atonement*, which was published by the University of Regina Press in 2016. Trevor's work has been nominated for numerous awards, including the Governor General's Award for Non-Fiction and the Writer's Trust Award for Non-Fiction. Trevor's writing has also appeared in the *Globe & Mail* and *Canadian Geographic*, as well as several anthologies.

Dr. Branimir Gjetvaj is a biologist, photography instructor and internationally published environmental photographer specializing in natural history and western Canadian landscapes. He has participated in numerous nature conservation initiatives and frequently contributes his photographic skills to local environmental organizations. One of his photography projects culminated in the award-winning book *The Great Sand Hills: A Prairie Oasis*. In 2013 Branimir was recognized by the Canadian Environmental Law Association for extensive participation in several key environmental NGOs, and for using his photography to advance environmental conservation. He is the current President (2016-2018) of Nature Saskatchewan, a provincial conservation organization.